Elementary School Mathematics for Parents and Teachers

Volume 2

Elementary School Mathematics for Parents and Teachers

Volume 2

Raz Kupferman
The Hebrew University, Israel

Illustrations by **Philip Pekar**

World Scientific

NEW JERSEY • LONDON • SINGAPORE • BEIJING • SHANGHAI • HONG KONG • TAIPEI • CHENNAI • TOKYO

Published by

World Scientific Publishing Co. Pte. Ltd.
5 Toh Tuck Link, Singapore 596224
USA office: 27 Warren Street, Suite 401-402, Hackensack, NJ 07601
UK office: 57 Shelton Street, Covent Garden, London WC2H 9HE

Library of Congress Cataloging-in-Publication Data
Names: Kupferman, Raz.
Title: Elementary school mathematics for parents and teachers / by Raz Kupferman (The Hebrew University, Israel).
Description: New Jersey : World Scientific, 2015– | Includes index.
Identifiers: LCCN 2015030122 | ISBN 9789814699907 (hardcover : alk. paper : v. 1) | ISBN 9789814699914 (pbk. : alk. paper : v. 1)
Subjects: LCSH: Education--Parent participation. | Mathematics--Study and teaching (Elementary)
Classification: LCC QA135.6 .K825 2015 | DDC 372.7--dc23
LC record available at http://lccn.loc.gov/2015030122

Volume 2
ISBN 978-981-3108-92-9
ISBN 978-981-3108-93-6 (pbk)

British Library Cataloguing-in-Publication Data
A catalogue record for this book is available from the British Library.

Illustrations by Philip Pekar

Preface

How did I get to write this book?

I am a professor of mathematics at the Hebrew University of Jerusalem in Israel. In 2008, I inaugurated, together with a group of vigorous graduate students, a professional development program (PD) in mathematics for elementary school teachers. The purpose of our program was to enhance the content knowledge (CK) of elementary school teachers in Israel, as most of them lack post-scholar mathematical education. The uniqueness of our PD was that all the instructors were active mathematicians. This created a fascinating encounter between two groups of people—mathematicians and elementary school teachers—who have very different perspectives on what mathematics is about and how it should be taught.

One of my first revelations, when we started to plan the PD, was that there does not exist a single mathematics textbook (in Hebrew) for teachers. Teachers don't have any resource to compensate for their lack of content knowledge. This lack of resources also applies to resources of pedagogical content knowledge (PCK), a term coined by Shulman for content knowledge that is particularly relevant for teachers.

The lack of resources is particularly startling from the perspective of a university professor. I cannot imagine my work as a lecturer without an abounding literature, comprising hundreds of textbooks and reference books on any subject I might ever teach. I have benefited from such dedicated literature at all stages of my teaching: in consolidating my own knowledge, in building a syllabus and in constructing lesson plans. These resources have provided me with exposure to perspectives of other lecturers as well as examples and exercises. Such resources, I discovered, are not available to the thousands of mathematics teachers in Israel.

While researchers unanimously agree on the strong correlation between CK and high quality teaching, the very idea that teachers should gain expertise in elementary school mathematics is foreign to the general public. Everybody understands that teachers need to be trained in order to transmit their knowledge on to their students, but only a few believe that teachers need to consolidate their own knowledge of the subject matter. Elementary school mathematics is usually perceived as rudimentary common knowledge. "Aren't teachers proficient in arithmetic?", I was asked more than once, or, "Can't teachers distinguish a triangle from a square?".

CK and PCK involve much more than "distinguishing a triangle from a square". Being acquainted with the curriculum is certainly a prerequisite for teachers, but the more important factor is the depth of their knowledge. Mathematics is not just an assortment of facts to learn and skills to practice. Mathematics is perhaps the climax of human intelligence: a science in which abstraction and generalization are central concepts; a science characterized by recurring patterns and inter-relations; a science that offers an inexhaustible collection of fascinating riddles for all ages and levels.

Good teachers must have an in-depth knowledge of mathematics in order to pass on the essence of this science and the joy of its practice to their students. High quality teaching requires the ability to distinguish between the essential and the inessential, the general and the specific, as well as to be able explain a concept in several ways, to find the most powerful metaphors and forms of representation, to provoke critical thinking and self-analysis, to choose good examples and exercises, and to rapidly and accurately identify students' errors and misconceptions. These diverse abilities don't appear out of the blue. They have to be acquired and honed. One way of doing this is through dedicated professional literature.

While I was preparing my own lesson plans for the PD, I started to write down some notes on elementary school mathematics, which, I hoped, would be valuable for teachers. Three years later, I collected this assortment of lecture notes and lesson plans into a textbook on elementary school mathematics.

Who should read this book?

When I started to write this book, my target audience was elementary school teachers. In the course of writing I became aware of a much larger group of people who could benefit from this book—parents of elementary school children. Parental involvement in their children's education has been on the rise for many years: at home, whether through home schooling or in public education, through parent-teacher activity, and more. With this in mind, I decided to address this book to everyone involved in children's mathematical education, whether as a certified teacher or as a parent. Henceforth, whenever I refer to **teachers** or **educators**, these terms should be interpreted in the broadest sense: every parent is automatically an educator!

A third category of target readers consists of people who wish to rediscover the mathematics they studied at a young age from a more mature perspective. In summary: this is a book for adults on children's mathematics.

The benefits that readers are expected to gain are the following:

(1) **Content knowledge**: This book is foremost a treatise on elementary school

mathematics. It exposes the reader to the full school curriculum in a logical and systematic order. It provides the reader with a coherent exposition that integrates the various topics, thus presenting mathematics as a whole. This unified view, combined with a meticulous attention to details, provides educators with the depth of knowledge required by their profession.

(2) **Pedagogic content knowledge**: While PCK is commonly contrasted to "pure" CK, the distinction between the two is often fuzzy. For example, the order of exposition, and the choices made all along this book as to which topics to expand on versus which topics to reduce, constitute important PCK.

In addition, this book offers the reader traditional PCK: Facts and properties are exposed and explained in more than one way; definitions, theorems and proofs are adapted to fit early stages of math education; distinctions between "easy" and "hard" are stressed and analyzed; the interplay between the concrete and the abstract is emphasized; recurring errors and suggestions on how to eliminate them are presented; relations between the current topic and previously studied topics as well as material to be taught are presented, thus seeding the foundations for future learning; instructive metaphors and useful means of representations are incorporated in every subject; examples, exercises, and enjoyable activities are suggested.

(3) **Get a glimpse on how mathematicians think**: While some may wonder whether this is a blessing or a curse, I believe that educators have much to benefit from a mathematician's exposition of elementary school mathematics. I have always believed the distinction between elementary and advanced mathematics to be artificial and inappropriate. In a sense, the more (seemingly) elementary mathematics is, the closer it is to the most basic human ways of reasoning, abstracting, and generalizing. These three skills are the crux of a mathematician's work, and they constitute the main objective of math education.

(4) **Activities**: CK and PCK are for educators. Children, especially young children, learn through concrete riddles and problems. It is through repeated experience that their personal perception of mathematics develops. Throughout this book, I have included explicit suggestions for activities or games that can be played either at home or in class.

(5) **Fun**: As the reader must have guessed by now, I view mathematics as a pinnacle of human inventions. While mathematics is usually advocated for its usefulness, its beauty is, in my opinion, downplayed. I believe this book can provide many adults with the corrective experience that they deserve. Those, who as children, strived to survive mathematics, at last have the opportunity to enjoy it.

What is in this book?

This book is the second of three volumes. The three volumes cover the elementary school math curriculum common in most parts of the world. The first volume focuses on the early years (roughly Grades K–2). The second volume focuses on the subsequent years (roughly Grades 3–6). The third volume covers some remaining K–6 topics, as well as topics on the border of elementary school and middle school (e.g., negative numbers and pre-algebra). This classification into grade levels should only be viewed as a rough estimate, not only because curricula vary between countries, but primarily because the same subject may be taught in different grades, at an increasingly advanced level of sophistication.

The present volume is divided into two main parts:

(a) Chapters 19–26 are concerned with whole number arithmetic. Chapter 19 deals with multiplication algorithms, whereas Chapter 20 treats division algorithms. Chapter 21 is devoted to the evaluation of composite arithmetic operations, presenting the rules of precedence and the use of parentheses (known as *the order of operations*). Chapter 22 is devoted to division with a remainder. Chapter 23 introduces the notions of multipliers and divisors, as well as prime numbers, which can be viewed as the building blocks of all natural numbers. Chapter 24 considers common multipliers and common divisors of given numbers; aside from being interesting on their own, finding common multipliers and common divisors is a prerequisite for some operations with fractions. Chapter 25 covers the topic of divisibility rules, which are easy-to-apply criteria testing whether numbers are divisible by a given divisor. Chapter 26 considers number sequences.

(b) Chapters 27–32 are concerned with fractions. Chapter 27 introduces fractions, focusing on their role in quantifying parts of a whole. Chapter 28 shows that fractions represent quotients of natural numbers, and introduces the fact that different fractions may have the same value. Chapter 29 deals with comparison between fractions, from concrete methods to algorithmic methods based on common denominators. The last three chapters focus on arithmetic operations with fractions: addition and subtraction in Chapter 30, multiplication in Chapter 31 and division in Chapter 32.

The introduction of every chapter contains three recurring sections, entitled **Hurdles**, **Remedies** and **Opportunities**. The Hurdles section describes difficulties and misconceptions that children may encounter in the context of that chapter. The Remedies section offers guidelines for overcoming those difficulties, or tips for preventing those difficulties from occurring. The Opportunities section suggests ways in which the current topic can be harnessed to "seed ahead" toward content appearing later in the curriculum.

Every chapter ends with a set of exercises. These exercises are multi-purpose: some exemplify topics learned in the chapter; some challenge the reader with teaching dilemmas;

some present typical errors made by children, and ask the reader to suggest possible ways to intervene. Keep in mind that this book is addressed to adults, and the exercise sets are as well. Yet, some of the exercises can serve, perhaps after some adaptation, as challenges for advanced children.

The book has two indices. The first is a "standard" index—an alphabetic list of keywords. The second index is sorted by tags from the American Common Core Standard.

Pedagogical principles

In a textbook of this type, there can be no sharp distinction between mathematics and pedagogy. Even though this is a book for adults, the writing and the content were strongly influenced by my perception on how mathematics ought to be presented to children. In this section I spell out some of the principles I believe should prevail in math education, especially for young children. These principles should serve as guidelines whenever considering transferring contents from this book to children.

(1) **Unpacking**: One of the most important challenges in math teaching is to decompose mathematical concepts into elementary constituents. Educators must be aware of all, sometimes subtle, dependencies between various topics. If Subject A is built upon Subjects B and C, then B and C have to be taught before A, and in a manner that paves the way to the later assimilation of A. For example, arithmetic relies on the number concept and counting. Teachers of the lower grades must have this in mind at all times, and make sure that their students internalize all aspects of numbers and counting that are relevant to arithmetic at the proper time.

(2) **Concrete and abstract**: Mathematics is the art of abstraction, however abstraction is not a topic in mathematics; it is a process that occurs in an individual's mind. Moreover, any abstraction must relate to something—to the concrete entities it is supposed to represent. Even though the ultimate goal of math teaching is to endow the learner with the powers of abstract reasoning, abstraction cannot be taught per se. Most of the learning process must occur on the concrete level. It is through a profusion of concrete mathematical experiences, that children gradually develop the ability to draw general conclusions. Please note that something abstract for one person may well be concrete for another. Manipulating numbers is abstract for a kindergartner, who would benefit more from manipulating and counting material objects, while it is concrete to a highschooler, compared to manipulating algebraic expressions.

(3) **Learning by doing**: "I hear and I forget; I see and I remember; I do and I understand" is a proverb attributed to Confucius. This short adage encapsulates an uncontested truth: active learning is more effective than passive learning. When it comes to children (in fact, also when it comes to adults), embarking

into an activity is called playing, and luckily, children love to play. This presents educators with the challenge of creating engaging, enjoyable activities, aimed at attaining specific pedagogic goals.

(4) **No nonsense**: Most textbooks and computer applications tend to sugarcoat their content, believing that it makes math more attractive to children. Thus, children find themselves playing with number-eating cows, popping clouds labeled by numbers, and so on. The problem with sugarcoating is that it conveys the message that math is not sufficiently attractive in its own right. This is a self-fulfilling prophecy. Math can be made fascinating without decorating it with infantile metaphors. While the boundaries between the playful and the infantile are sometimes fuzzy, and are a matter of taste, there is one thing I am sure of: if educators don't find the mathematical content interesting, this message will be transferred to children.

(5) **Challenging**: One of the unfortunate realizations I came up with through my encounter with teachers, was their fear of challenging children. "We should not confuse our students" is a warning I have heard too often. The teachers' fears, of course, do not come out of the blue. They are deep rooted in a general atmosphere that tends to over-protect children from failure.

To me, math without initial confusion is an oxymoron. The whole essence of math is to disentangle mind-boggling riddles with the power of our brain. Moreover, I don't believe math can be fun without a challenge. Children don't buy the "good job" if they feel that the task they have just fulfilled requires no effort. What needs to be changed is the attitude toward errors. Errors do not mean failure; errors are a consequence of trying and not giving up. There can be no meaningful math without unequivocally legitimizing making errors.

Acknowledgments

I am indebted to a great number of people, who helped me shape this book since its very early drafts. Foremost, I would like to thank the wonderful team of young mathematicians who joined me in inaugurating and carrying out the PD: Menny Aka, Shaull Almagor, Eran Asaf, Ayala Byron, Noa Eidelstein, Chaim Even Zohar, Ariel Goldstein, Yannai Gonczarowski, Shlomi Hod, Or Landsberg, Noa Nitzan, Ori Parzanchevsky, Shir Peled, Alon Pinto, Doron Puder, Shani Regev, Ori Rosenstein, Ron Rosenthal, Roy Schwarz, Moria Sigron, Lior Yanovski and Amitai Zernik. We were all closely monitored by Jason Copper and mentored by Abraham Arcavi, who was the one to first suggest the writing of this book.

The PD would have not become a reality without the massive cooperation of the Israeli Ministry of Education. For this I am indebted to the Superintendent of Mathematics Education, Hannah Perl, the national counselor Tami Giron, and the municipal counselors, Nava Atad, Cilla Rott and Tami Razvag.

I am happy to mention two books that served as inspiration: *A Third Divided by a Quarter* by Mira Ofran, and *Arithmetic for Parents* by Ron Aharoni (both published in Hebrew).

It is hard to imagine this book without the contribution of those who have read it, and provided me with invaluable comments, criticism, and constructive suggestions: Abraham Arcavi, Eran London, Ron Rosenthal, Yossi Shamai, Shimon Schocken and Anna Zarkh.

This book was originally written in Hebrew. Gal Kusturica was the editor of the Hebrew edition, to whom I owe the transformation of this book from a pile of notes to a (hopefully) coherent and accessible textbook. I am grateful to Moshe Devere for the English translation and to Sharon Hirsch for her meticulous copy-editing.

Illustrations and graphics play a very important role in books of that type. I am grateful to Philip Pekar for the illustrations, and to Alex Bernstein for useful graphic advice.

Finally, I would like to express my appreciation to the Hebrew University, for its commitment to education, and for granting us all with the academic freedom to apply our knowledge wherever we believe it can be useful.

Contents

Chapter 19

Multiplication of Multi-Digit Numbers

In Chapter 7, we introduced the multiplication operation. The focus in that chapter was on getting acquainted with various interpretations of this operation, understanding its properties, and gaining proficiency in evaluating products of single-digit numbers (i.e., evaluating products within the range of the commonly used 10×10 **multiplication table**). In this chapter we will expand our proficiency by learning to evaluate products of arbitrarily large numbers. We will illustrate several alternative methods, including two algorithms taught in school: the **expanded algorithm** and the **standard algorithm**.

Hurdles:

(a) Children struggle multiplying large numbers while not mastering the multiplication table.
(b) Children have a hard time memorizing the multiplication algorithms, and as a result commit errors.
(c) Children adopt a how-to attitude; they get into the unhealthy habit of acting mechanically without looking for a meaning.
(d) Children don't have means to validate their solutions.

Remedies:

(a) Keep practicing products of single-digit numbers until full mastery has been achieved.
(b) Practice products of one-significant-digit numbers, such as 7×80, 60×40 and 500×20.
(c) Teach multiplication algorithms gradually: Start with concrete demonstrations of the distributive property; proceed with the slightly more abstract area model and grid model; connect the concrete representations to the expanded algorithm;

finally, transition gradually from the expanded algorithm to the standard algorithm. Be patient—conceptual understanding is more valuable than mechanical mastery.

(d) Get children into the habit of validating products using estimates; recall that estimates are sometimes as valuable as a precise answer.

(e) Encourage mental calculations; they may help internalizing the evaluation algorithms while building up self-esteem.

Opportunities:

(a) Exploit multiplication algorithms to reinforce in children's minds the principles underlying the decimal representation system.

(b) Get acquainted with alternative evaluation algorithms; these offer glimpses on fascinating chapters of mankind history and reveal mathematical gems that are not part of standard curricula.

19.1 Repeated addition

Before learning algorithms for evaluating products of multi-digit numbers, we should mention that products of whole numbers can be evaluated, in principle, using **repeated addition**. For example, 12×536 can be evaluated by appending 536 repeatedly 12 times:

$$\begin{aligned} 536 + 536 &= 1072 && \text{(2 times 536)} \\ 1072 + 536 &= 1608 && \text{(3 times 536)} \\ 1608 + 536 &= 2144 && \text{(4 times 536)} \\ \vdots \quad & \quad \vdots && \\ 5896 + 536 &= 6432 && \text{(12 times 536)} \end{aligned}$$

Clearly, this approach is inefficient: the number of sums that need to be evaluated is as large as the size of the multiplier. Just imagine the amount of work that would be involved in multiplying, say, two four-digit numbers using repeated addition.

Before teaching children algorithmic methods for evaluating products, it is dearly recommended to challenge them with tasks that encourage developing intuitive multiplication strategies.

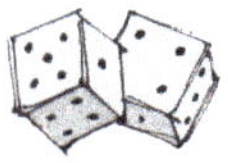

Activity: Products beyond the multiplication table

Consider the following word problem:

> A deck of cards has 4 suits: clubs, spades, hearts, and diamonds. Each suite comprises 13 cards (ace to 10, jack, queen, and king). What is the total number of cards in a deck?

First, children have to realize that the solution to this word problem is the value of a product, 4×13. To evaluate this product, they can invent a variety of strategies. For example:

(1) Skip-count by intervals of 4 on the hundred table (Volume 1, p. 21).
(2) Skip-count by intervals of 13 on a number line.
(3) Use the **distributive property** of multiplication: add 10×4 and 3×4.
(4) Use the **associative property** of multiplication: 4×13 is twice as much as 2×13.

These strategies are only a fraction of a variety of methods that children might develop. If children react to the challenge positively, you can raise the level of difficulty gradually up to products of two-digit numbers.

Such challenges sharpens children's intuition, and reinforce their perception of the multiplication operation. Furthermore, children's attempts to invent solution strategies of their own often reveal misconceptions, and offer an excellent opportunity to treat them at an early stage. For example, a common error resulting from an erroneous interpretation of the distributive property leads children to believe that

$$12 \times 13 \quad \text{equals} \quad 10 \times 10 + 2 \times 3,$$

namely, 106. If a child commits such an error, one could ask whether they expect the product to be greater or less than 10 times 13, and thus lead them toward the correct interpretation of the distributive law.

19.2 Products of one-significant-digit numbers

Suppose we were proficient evaluating products of single-digit numbers. Thus, we would know by heart the following multiplication fact:

$$7 \times 8 = 56.$$

Could we use this knowledge to evaluate the following multi-digit products:

$$70 \times 8$$
$$7 \times 80$$
$$70 \times 80$$
$$700 \times 8$$
$$7000 \times 80.$$

These five products have a common property. Their factors are characterized by having only one digit that is not zero—their left-most digit, also known as the **most significant digit**. Thus, such numbers are called **one-significant-digit numbers**. The sequence of one-significant-digit numbers starts with

$$1, 2, 3, \ldots, 9, 10, 20, 30, \ldots, 90, 100, 200, \ldots, 900, 1000, 2000, \ldots.$$

One-significant-digit numbers are either single-digit numbers, or a product obtained by multiplying a single-digit number with a **power of ten**. For example,

$$70 = 7 \times 10$$
$$700 = 7 \times 10 \times 10$$
$$7000 = 7 \times 10 \times 10 \times 10,$$

and so on.

Equipped with this new terminology—one-significant-digit numbers—let's revisit the five products shown above. In all cases, the two factors are one-significant-digit numbers, the significant digits being 7 and 8, respectively. To evaluate those products, we first express each factor as a product of a single-digit number and a power of ten, and then apply both the commutative and the associative properties of multiplication (Volume 1, pp. 82, 84) to reorder the products as follows:

$$70 \times 8 = (7 \times 10) \times 8 = (7 \times 8) \times 10$$
$$7 \times 80 = 7 \times (8 \times 10) = (7 \times 8) \times 10$$
$$70 \times 80 = (7 \times 10) \times (8 \times 10) = (7 \times 8) \times 10 \times 10$$
$$700 \times 8 = (7 \times 10 \times 10) \times 8 = (7 \times 8) \times 10 \times 10$$
$$7000 \times 80 = (7 \times 10 \times 10 \times 10) \times (8 \times 10) = (7 \times 8) \times 10 \times 10 \times 10 \times 10.$$

Examine the right-hand sides: they consist of a chain of products, starting with 7×8, and followed by one or more multiplications by ten. Recalling that $7 \times 8 = 56$, and that multiplication by ten results in a zero being appended to the right of the other factor (Volume 1, p. 94), we obtain

$$70 \times 8 = 560$$
$$7 \times 80 = 560$$
$$70 \times 80 = 5600$$
$$700 \times 8 = 5600$$
$$7000 \times 80 = 560000,$$

where to help keep track of the "zeros", we colored those zeros originating from the first factor orange, and colored those originating from the second factor blue.

These examples suggest the following systematic procedure: the product of one-significant-digit numbers can be evaluated in two stages:

(1) **Disregard the zeros**: Evaluate the product of the most-significant digits of the two factors.
(2) **Append the zeros**: Take the result of the first stage, and append to its right as many zeros as were disregarded.

The following figure demonstrates how to apply this algorithm:

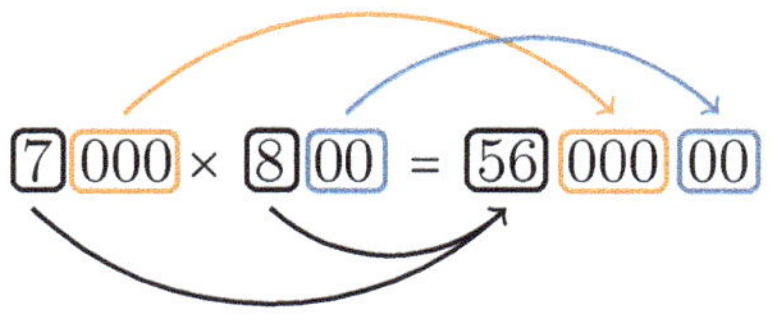

Please note: The evaluation of products of one-significant-digit numbers is a prerequisite for evaluating products of multi-digit numbers. It is therefore essential to gain high proficiency of this type of multiplication before proceeding to the multiplication of multi-digit numbers.

Sneak preview: products of decimals

There are many similarities between products of multi-digit numbers and product of decimals. In the latter case, one-significant-digit decimals are of the form,

$$0.1, 0, 2, \ldots, 0.9, 0.01, 0.02, \ldots, 0.09, 0.001, 0.002, \ldots.$$

They are all products of single-digit numbers and powers of 0.1.

Products of one-significant-digit decimals are evaluated very similarly to products of one-significant-digit whole numbers, as in the following example:

$$\begin{aligned} 0.03 \times 0.004 &= (3 \times 0.1 \times 0.1) \times (4 \times 0.1 \times 0.1 \times 0.1) \\ &= (3 \times 4) \times (0.1 \times 0.1 \times 0.1 \times 0.1 \times 0.1) \\ &= 0.00012. \end{aligned}$$

19.3 Products of a single-digit number and a multi-digit number

A useful strategy in mathematics is to solve a problem in stages. Before considering products of two multi-digit numbers, we will develop the required tools by considering the somewhat simpler case, where one of the factors is a single-digit number.

19.3.1 Using the distributive property

19.3.1.1 Example 1

Consider the following question:

> Ms. Merz's class prepared 7 gift packages. Each package contained 13 muffins. What was the total number of muffins in all the gift packages?

This question presents a typical multiplicative scenario. A fixed number of items—13 muffins—repeats itself a given number of times—7 times. The aim is to calculate the total number of items, which is given by the following product,

$$7 \times 13.$$

It is possible to evaluate this product by the repeated addition of 13, but as previously noted, this approach is not efficient, and becomes impractical with larger numbers.

A more efficient approach is based on the distributive property of multiplication (Volume 1, p. 86). We partition the multiplicand, which is a two-digit number—13—into a sum of tens and ones—$10 + 3$. Then, we apply the distributive property:

$$7 \times 13 = 7 \times (10 + 3) = (7 \times 10) + (7 \times 3).$$

Examine the right-hand side: it exhibits a sum of *two* products. Seemingly, the use of the distributive property has only complicated the task, which was the evaluation of just one product. The crux of the matter, however, is that each one of the two products involves only one-significant-digit numbers, and is therefore easier to evaluate. Thus,

$$7 \times 13 = \underbrace{(7 \times 10)}_{70} + \underbrace{(7 \times 3)}_{21} = 91.$$

If these arithmetic equalities are too abstract for a child, it is always possible to replace them by concrete equivalents. For example:

> We wish to put 13 muffins in each of the 7 gift packages. At a first stage, we put 10 muffins in each package. (How many muffins are needed?) At the second stage, we add 3 muffins to each package. (How many additional muffins are needed?)

The distributive property, as applied to the present context, is illustrated below.

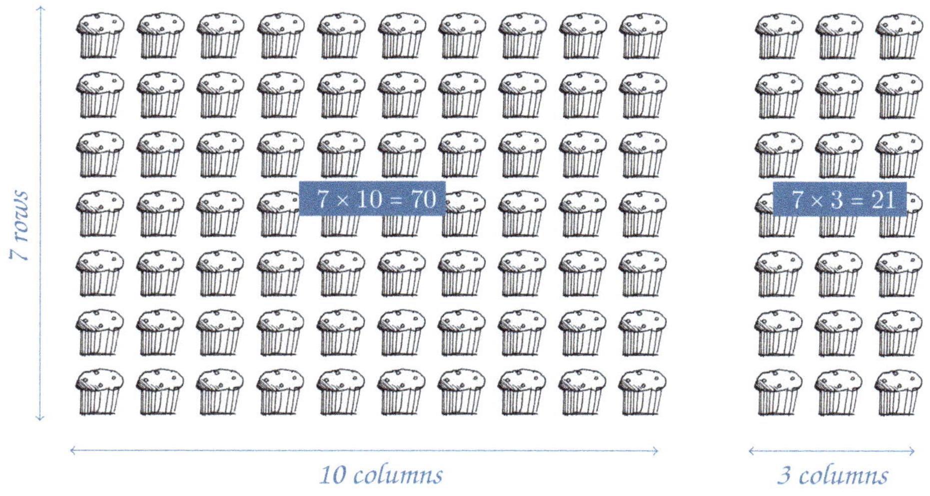

Such visual aids can be very useful for disentangling and elucidating a mathematical problem. As a rule, visual aids should be kept as simple as possible, and should be easy for a child to reproduce. Obviously, drawing 91 muffins does not fulfill this requirement.

An alternative concretization of the distributive property is depicted below:

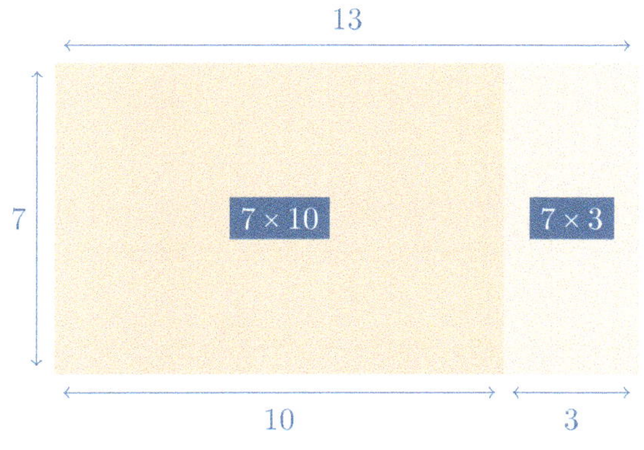

This illustration is based on the **area** concept, and its relation to multiplication (Volume 1, p. 232). The area (in square units) of a rectangle is the product of the lengths (in length units) of a pair of adjacent sides. The product 7×13 can therefore represent the area of a rectangle whose sides are 7 and 13 units long. Since area is additive, we can calculate the rectangle's area by partitioning it into two sub-rectangles, and adding up their areas. We partition the 13-unit-long sides into 10 and 3-unit-long segments. As with the muffins, the evaluation of the areas of the sub-rectangles involves only products of one-significant-digit numbers, which is a skill that we have already acquired.

Please note: since the figure is only used as an illustration, it is not important to maintain accurate proportions. In fact, accurate proportions would be impractical beyond two-digit numbers. No confusion should occur if the *intended* length of each segment is clearly written, as we did in the figure above.

19.3.1.2 Example 2

Consider the following product:

$$2 \times 324.$$

This time, the multiplicand is a three-digit number. Therefore, we can partition it into a sum of hundreds, tens, and ones. Applying the distributive property,

$$2 \times 324 = 2 \times (300 + 20 + 4) = (2 \times 300) + (2 \times 20) + (2 \times 4).$$

An illustration using the area model is depicted below:

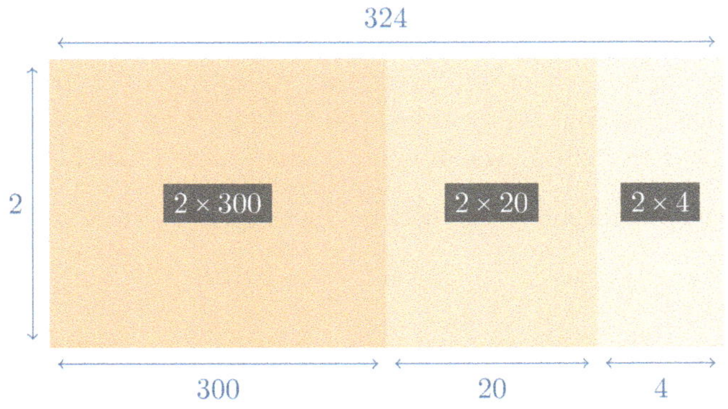

The rectangle, whose sides are 324 and 2 units long, is partitioned into three sub-rectangles. The area of each sub-rectangle is the product of one-significant-digit numbers, which we can readily evaluate:

$$2 \times 324 = \underbrace{(2 \times 300)}_{600} + \underbrace{(2 \times 20)}_{40} + \underbrace{(2 \times 4)}_{8} = \boxed{648}.$$

Another common way of keeping track of the same solution strategy is using a table (a method also known as the **grid model**). Partition the multi-digit factor into a sum of hundreds, tens and ones, and place each part—which is a one-significant-digit number—at the head of a column. Place the other factor, which is a single-digit number, at the head of a row, as shown below:

×	300	20	4
2			

Next, fill-in the table. Place in each cell the product of the number heading that row and the number heading that column:

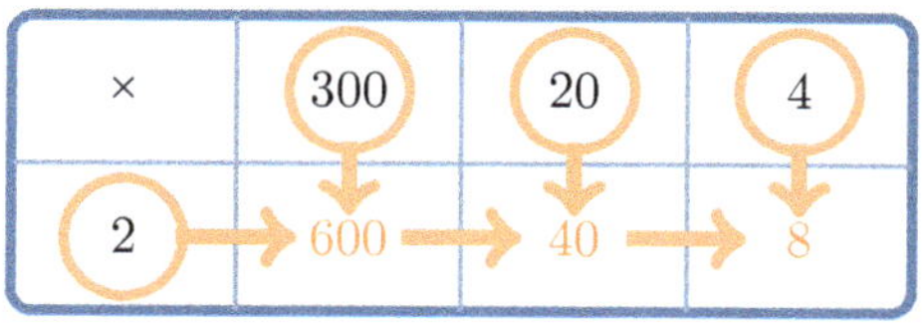

Finally, add together the three products, yielding once again

$$2 \times 324 = 600 + 40 + 8 = \boxed{648}.$$

19.3.1.3 Example 3

Consider yet another product:

$$4 \times 637.$$

As shown above, we can partition the multiplicand into hundreds, tens and ones,

$$637 = 600 + 30 + 7,$$

and apply the distributive property,

$$4 \times (600 + 30 + 7) = (4 \times 600) + (4 \times 30) + (4 \times 7).$$

Once again, we have reduced the problem into evaluating products of one-significant-digit numbers and adding them up. Using the grid model:

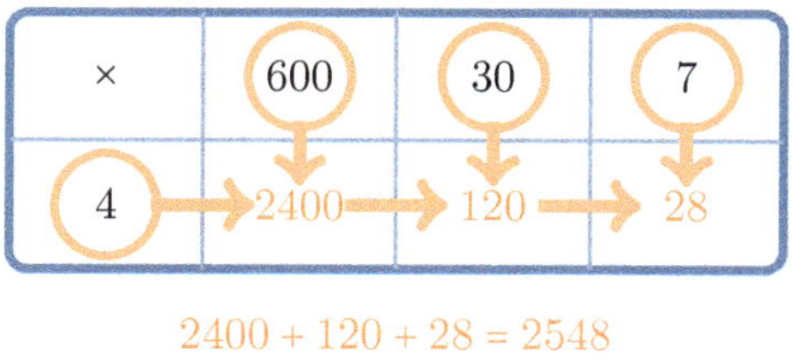

The distributive property of multiplication, as exemplified in the present section, is at the heart of the multiplication algorithms presented in the next sections. I strongly recommend to precede the instruction of algorithms with a prolonged acquaintance with the methods and concretizations presented in this section. This hands-on approach helps clarify the principles upon which the algorithms are based. Otherwise, algorithms remain obscure, technical procedures, and the lack of understanding results in errors and frustration.

19.3.2 Multiplication without regrouping

Illustrations play an important role in the learning process. Eventually, though, one needs evaluation methods that are methodical, efficient and performable on the back on an envelope. This is what algorithms were invented for. In this section, we will present two multiplication algorithms–the expanded algorithm and the standard algorithm. Both algorithms are essentially the same, and they only differ in how intermediate calculations are recorded. The expanded algorithm, as its name indicates, records intermediate calculations in greater detail, and is therefore easier to remember and apply.

For didactic purposes, we will consider the same collection of products that were evaluated in the previous section by concrete means. Consider first the product:

$$2 \times 324.$$

It is convenient to perform arithmetic algorithms on grid paper. The first step is to write down the two factors one above the other, so that digits representing the same decimal units are aligned vertically. Since multiplication is commutative (Volume 1, p. 82), the order of the factors is immaterial. It is conventional, though, to write the number that has more digits on top, as follows:

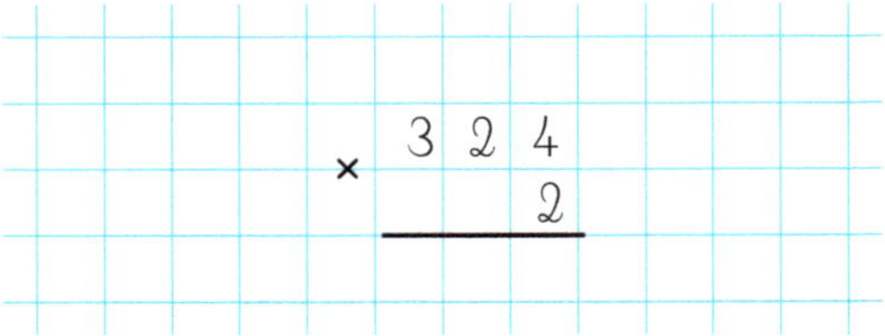

19.3.2.1 The expanded algorithm

Applying the **expanded algorithm**, we proceed as follows (hereafter, the labeled stages refer to the stages depicted in the figure below):

(a) **Multiply ones by ones**: Multiply the second factor by the ones of the first factor, in this case, 2×4. Write the product, 8, under the two factors, maintaining the right-alignment of all the numbers.

(b) **Multiply ones by tens**: Multiply the second factor by the tens of the first factor, in this case, 2×20, which is a product of one-significant-digit numbers. Write the product, 40, under the result of Step (a).

(c) **Multiply ones by hundreds**: Multiply the second factor by the hundreds of the first factor, in this case, 2×300, which is a product of one-significant-digit numbers. Write the product, 600, under the result of Step (b).

(d) **Add up the three products**: Add up the products obtained in Steps (a)–(c). The sum, which is the product we are looking for, is written, as usual, under the three addends.

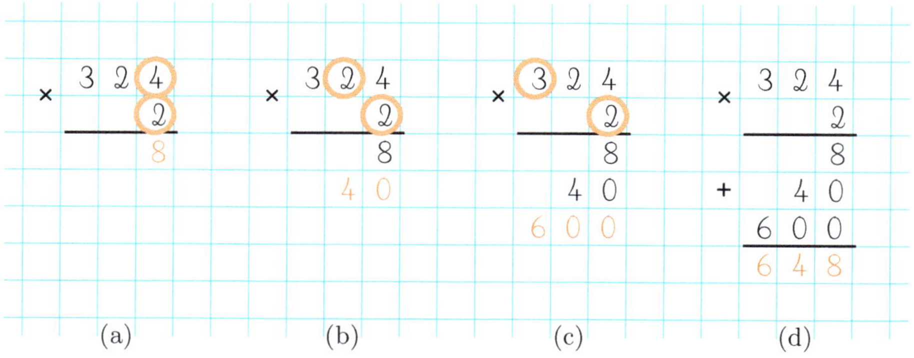

Compare this algorithm with the table on p. 9. The same three products are evaluated and added up. The only difference is that the operands, the intermediate results and the final product are written vertically, rather than being organized in a table. The expanded algorithm is a natural intermediary between a visual application of the distributive property and the standard algorithm, which is optimized for minimum writing.

19.3.2.2 The standard algorithm

The standard algorithm follows the same steps as the expanded algorithm, but writes all the results in a single line:

(a) **Multiply ones by ones**: 2 times 4 ones are 8 ones. Write the digit 8 in the ones column of the product, under the two factors.

(b) **Multiply ones by tens**: 2 times 2 tens are 4 tens. Write the digit 4 in the tens column of the product.

(c) **Multiply ones by hundreds**: 2 times 3 hundreds are 6 hundreds. Write the digit 6 in the hundreds column of the product.

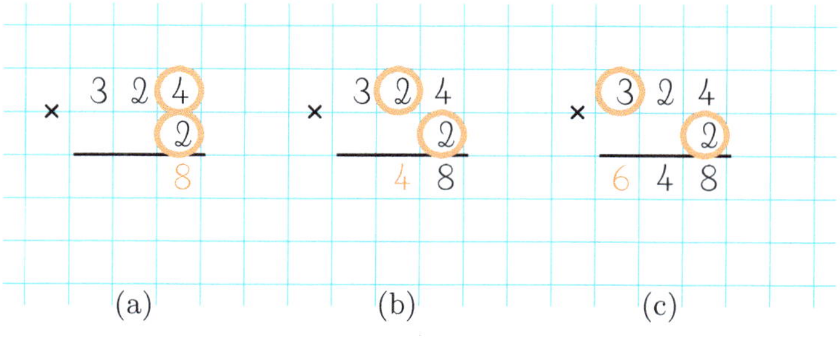

At this stage, the standard algorithm may seems simpler than the expanded algorithm. Indeed, we started with a very simple example, in which the clarity of the expanded algorithm is hard to discern. The distinction between the two algorithms will become apparent once regrouping is required.

19.3.3 Multiplication with regrouping

Consider the product,

$$4 \times 637.$$

Once again, we start by writing the two operands vertically, the multi-digit factor above the single-digit factor:

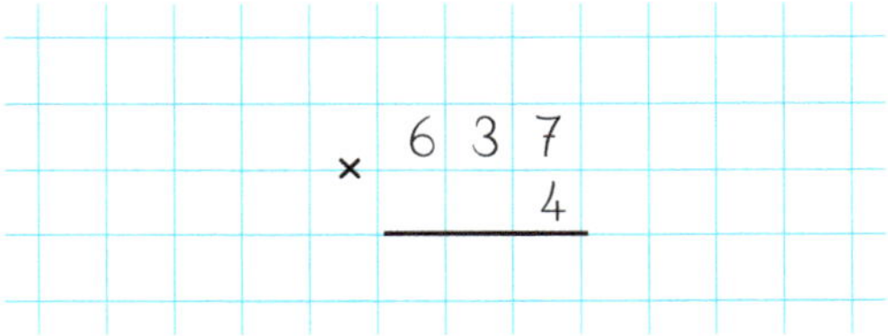

19.3.3.1 The expanded algorithm

The expanded algorithm proceeds as follows:

(a) **Multiply ones by ones**: Multiply the second factor by the ones of the first factor, in this case, 4×7. Write the product, 28, under the two factors, maintaining the right-alignment of all the numbers.

(b) **Multiply ones by tens**: Multiply the second factor by the tens of the first factor, in this case, 4×30, which is a product of one-significant-digit numbers. Write the product, 120, under the result of Step (a).

(c) **Multiply ones by hundreds**: Multiply the second factor by the hundreds of the first factor, in this case, 4×600, which is a product of one-significant-digit numbers. Write the product, 2400, under the result of Step (b).

(d) **Add the three products**: Add the products obtained in Steps (a)–(c). The sum, which is the product we are looking for, is written, as usual, under the three addends.

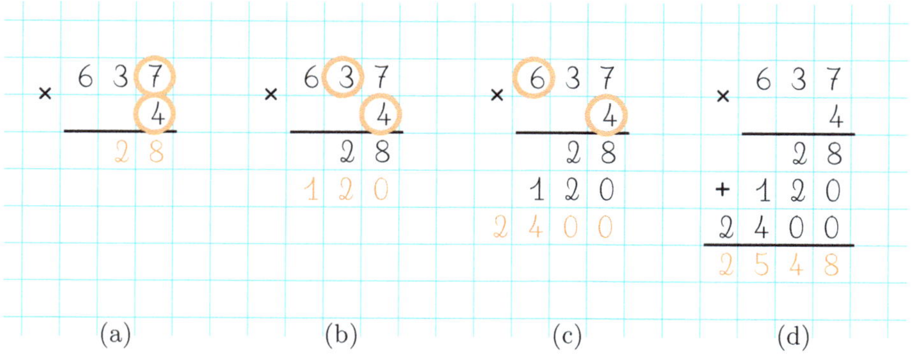

Compare the expanded algorithm with the table representation used on p. 9. Note the fundamental difference between this example and the previous one: In the present example, the product of ones comprises tens, the product of ones and tens comprises hundreds, and the product of ones and hundreds comprises thousands. That is to say, we need to regroup decimal units. This is not so apparent with the expanded algorithm, but becomes very explicit with the standard algorithm.

19.3.3.2 The standard algorithm

The standard algorithm proceeds as follows:

(a) **Multiply ones by ones**: 4 times 7 ones are 28 ones. Since we cannot write a two-digit number in the space allocated for the ones digit of the product, we regroup the 28 ones as 8 ones and 2 tens. We write the digit 8 in the space allocated for the ones digit of the product, and we write the digit 2 (the "carry") above the tens digit of the first factor, just as we did in vertical addition.

(b) **Multiply ones by tens**: 4 times 3 tens are 12 tens, to which we have to add the 2 tens carried from the previous step. Thus, we have a total of 14 tens, which we regroup as 4 tens and 1 hundred. We write the digit 4 in the space allocated for the tens digit of the product, and the digit 1 above the hundreds digit of the first factor.

(c) **Multiply ones by hundreds**: 4 times 6 hundreds are 24 hundreds, to which we have to add the 1 hundred carried from the previous step. Thus, we have a total of 25 hundreds, which we regroup as 5 hundreds and 2 thousands. We write the digit 5 in the space allocated for the hundreds digit of the product, and the digit 2 above the place allocated to the thousands digit of the first factor (which in the present example only counts three digits).

(d) **Add the thousands**: There are 2 thousands carried from the previous step. We write the digit 2 in the space allocated to the thousands digit of the product.

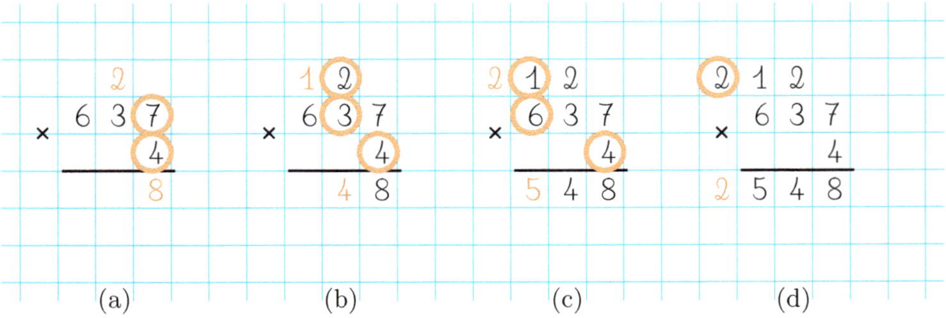

Notes

(1) Both the expanded and standard algorithms are based on the same underlying principles. In the expanded algorithm, the stages of multiplication and regrouping are separated ("first multiply, then regroup"), whereas in the standard algorithm they are interlaced.

(2) The standard algorithm is more succinct, but many children find it confusing and harder to implement. I recommend delaying using the standard algorithm until children have mastered the expanded algorithm. In fact, I have heard teachers affirm that they use the expanded algorithm almost exclusively, and resort to the standard algorithm only with advanced students.

19.4 Products of multi-digit numbers

After having learned to evaluate products of a multi-digit number and a single-digit number, we will harness this knowledge to evaluate products of two multi-digit numbers. We will use the following product as our running example:

$$73 \times 84.$$

19.4.1 Using the distributive property

As we did in the previous section, we decompose the second factor into a sum of tens and ones, $84 = 80 + 4$, and apply the distributive property of multiplication:

$$73 \times 84 = 73 \times (80 + 4) = (73 \times 80) + (73 \times 4).$$

The right-hand side is a sum of two products, in which the first factor is a two-digit number, 73, whereas the second factor is a one-significant-digit number. We may further simplify this expression by decomposing the first factor into tens and ones, $73 = 70 + 3$, and apply the distributive property once again:

$$73 \times 80 = (70 + 3) \times 80 = (70 \times 80) + (3 \times 80)$$

$$73 \times 4 = (70 + 3) \times 4 = (70 \times 4) + (3 \times 4).$$

Putting it all together:

$$73 \times 84 = (70 \times 80) + (3 \times 80) + (70 \times 4) + (3 \times 4).$$

We thus obtain a sum of four products, all involving nothing but one-significant-digit numbers, which we already know how to evaluate:

$$3 \times 4 = 12$$
$$3 \times 80 = 240$$
$$70 \times 4 = 280$$
$$70 \times 80 = 5600.$$

Adding up the four products:

$$73 \times 84 = 5600 + 240 + 280 + 12 = \boxed{6132}.$$

The repeated use of the distributive property is displayed below using the area model:

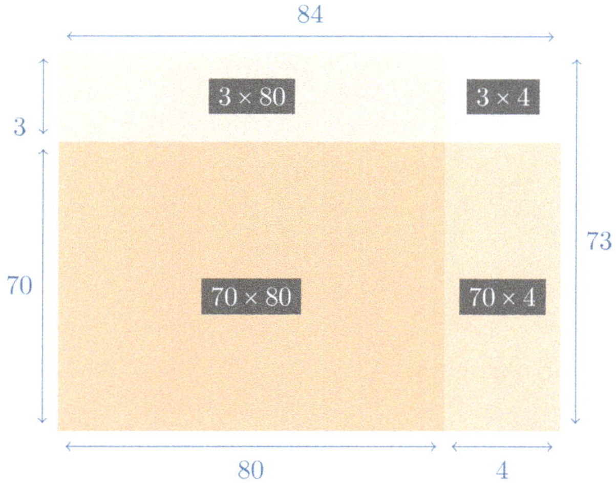

In this case, the original rectangle, whose dimensions are 73-by-84 length units is partitioned into four sub-rectangles, whose dimensions are one-significant-digit numbers. Once again, the partition does not have to be drawn to scale.

If we rather use the grid model, it comprises two rows and two columns.

×	80	4
70		
3		

We fill-in the table, placing in each cell the product of the number heading that row and the number heading that column, and finally add up the four products:

×	80	4
70	5600	280
3	240	12

$$5600 + 280 + 240 + 12 = 6132$$

19.4.2 Multiplication with regrouping

We start by writing the two factors one above the other. Since both factors are two-digit numbers, the order is arbitrary:

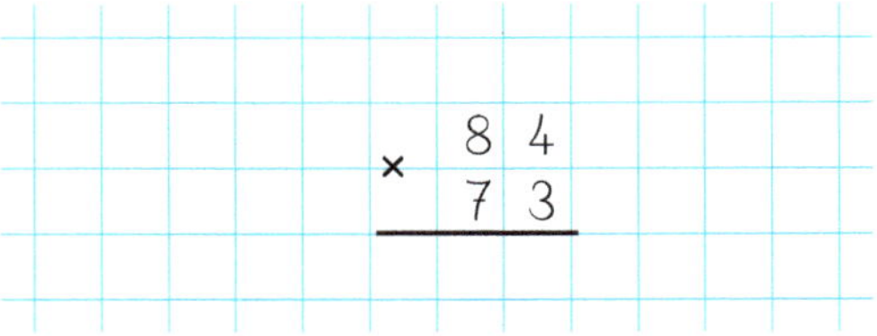

19.4.2.1 The expanded algorithm

The expanded algorithm proceeds as follows:

(a) **Multiply ones by ones**: Multiply the ones of the second factor by the ones of the first factor, in this case, 3×4. We write the product, 12, under the two factors, maintaining the right-alignment of all the numbers.

(b) **Multiply ones by tens**: Multiply the ones of the second factor by the tens of the first factor, in this case, 3×80. We write the product, 240, under the result of Step (a).

(c) **Multiply tens by ones**: Multiply the tens of the second factor by the ones of the first factor, in this case, 70×4. We write the product, 280, under the result of Step (b).

(d) **Multiply tens by tens**: Multiply the tens of the second factor by the tens of the first factor, in this case, 70×80. We write the product, 5600, under the result of Step (c).

(e) **Add the four products**: Add the products obtained in Steps (a)–(d). The sum, which is the product we are looking for, is written as usual under the four addends.

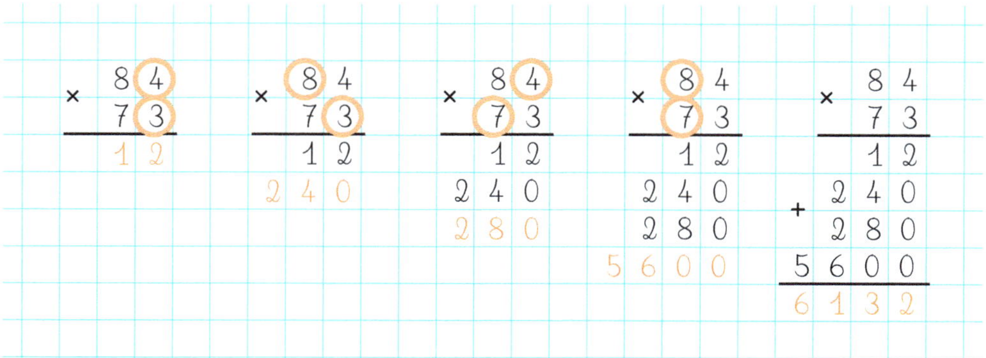

(a) (b) (c) (d) (e)

We stress again that all the intermediate calculations require only the evaluation of products of one-significant-digit numbers. Likewise, compare the expanded algorithm and the grid model on p. 16.

19.4.2.2 The standard algorithm

The standard algorithm for multiplying two multi-digit numbers is based on the following idea. Consider our running example, 73×84, and apply the distributive property after partitioning the first factor into a sum of ones and tens:

$$73 \times 84 = (3 + 70) \times 84 = (3 \times 84) + (70 \times 84).$$

Examine the first product, 3×84: one of its factors is a single-digit number, hence, we already know how to evaluate it using the standard algorithm:

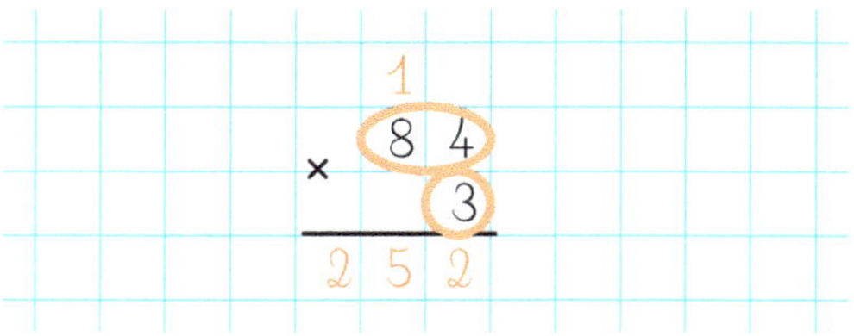

Now, consider the second product, 70×84: Since the multiplier is a one-significant-digit number, we can express it as a single-digit number times 10, and invoke the associative property of multiplication to reorder the product as follows:

$$70 \times 84 = (7 \times 10) \times 84 = (7 \times 84) \times 10.$$

We obtained a product, 7×84, in which one of the factors is a single-digit number, which is then multiplied by 10, i.e., shifted one column to the left. Thus, we evaluate 7×84 using the standard algorithm, but write each digit in the product one column to the left of where it would normally appear. To indicate that the product has been shifted, it is customary to place in the unused ones column a placeholder symbol, for example, an asterisk,

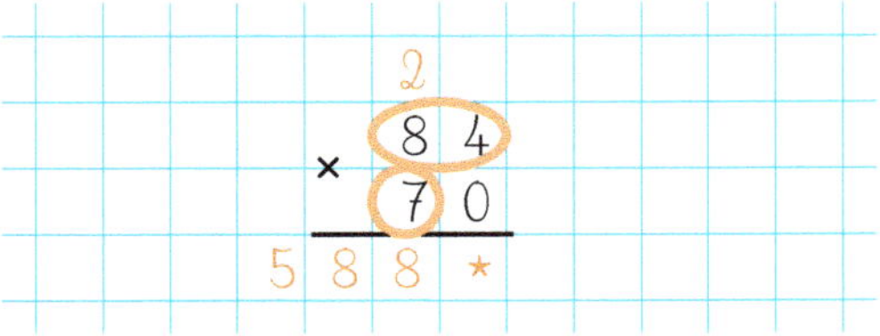

The standard algorithm combines these two stages as follows:

(a) **Multiply the first factor by the ones of the second factor**: Here we just repeat the above calculation.

(b) **Multiply the first factor by the tens of the second factor**: We repeat the last calculation, writing the shifted product under the result of Step (a). Note that the carry digits of Step (a) are no longer relevant, and should be ignored.

(c) **Add the two products**: Add the products obtained in Steps (a)–(b). The sum, which is the product we are looking for, is written as usual under the two addends.

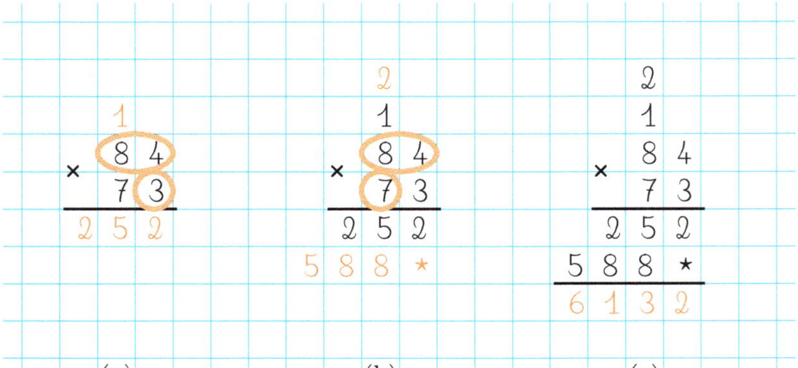

(a) (b) (c)

Please note: multiple regroupings result in multiple carry digits above the two factors. Sometimes, carry digits from two different stages are placed in the same column, causing confusion. I am not aware of any standard way for dealing with this problem. One possibility is to cross out a carry digit once it has been used. Another alternative is to arrange the carry digits vertically, as we did. In any case, do not ignore this technical difficulty, and adopt systematic measures to deal with it.

19.5 Mental calculations

Mental calculation is a term referring to arithmetic calculations performed in the brain, without using pen and paper. Many children find it difficult to even memorize the multiplication table. It is therefore not surprising that mental calculations beyond the multiplication table might daunt the struggling child. Nevertheless, I suggest encouraging children to develop gradually the capability of carrying out mental calculations. It is through mental calculations that children often develop a better sense for numbers, and internalize the properties of arithmetic operations. Moreover, mental calculations help develop evaluation skills—the ability to quickly obtain an estimate of the product.

The following list presents products of gradually increasing complexity. Such evaluations can serve as a guide to the development of mental calculation abilities:

(1) Start with the multiplication table, e.g., 6×9 and 7×8.
(2) Let one of the factors exceed 10, e.g., 11×7, 12×6 and 13×5. You can facilitate the task by presenting sequences of products, such as 10×6, 11×6, 12×6, etc.
(3) Practice products of one-significant-digit numbers, e.g., 4×30, 60×6 and 20×40.
(4) Proceed with products of two digit numbers and one-significant-digit numbers, e.g., 11×70 and 13×50. Here too, sequences of products might help, e.g., 13×5 followed by 13×50.
(5) Proceed to "easy" products of two-digit numbers, e.g., 11×12 and 21×13, after having evaluated 10×12 and 20×13.
(6) The courageous child may enjoy the challenge of full-fledged two-digit products, e.g., 78×53. The sky is the limit.

Do not rush things, and do not feel obliged to reach the most advanced level. Remember that the goal is to help children improve their number sense—not to perform as mathemagicians.

19.6 Other multiplication algorithms

This section is concerned with other multiplication algorithms. There are several good reasons for getting acquainted with algorithms that are not part of the formal curriculum. Sometimes, alternative algorithms are more efficient than the standard ones; on many occasions they reveal beauty and ingenuity; finally, they constitute an important part of general knowledge and provide a glimpse into the fascinating history of mathematics.

19.6.1 Napier bones

Napier's bones is a nickname for a calculator invented by 17th century Scottish mathematician, John Napier (also renown for having invented the logarithm). The calculator is a sort of multiplication table, whose cells are tiles. Each tile represents a product of one-digit numbers. Take for example the tile corresponding to the product 8×3: rather than simply writing its value 24 in the center of tile, the ones digit, 4, is written on the lower-right corner, whereas the tens digit, 2, is written on the upper-left corner. The two digits are separated by a diagonal line that runs from the upper-right corner to the lower-left corner:

The complete set of tiles is shown below:

The best way to present Napier's algorithm, is through an example: Consider a product of three-digit numbers:

$$725 \times 834.$$

(a) Form a table whose columns are headed by the digits of the first product—the ones digits at the right—and its rows are headed by the digits of the second product—the ones digits at the bottom:

(b) Place Napier tiles in the table's cells. Consider, for example, the upper-left cell, where the column is headed by the digit 7 and the row is headed by the digit 8. Place the tile corresponding to the product 7×8 in that cell.

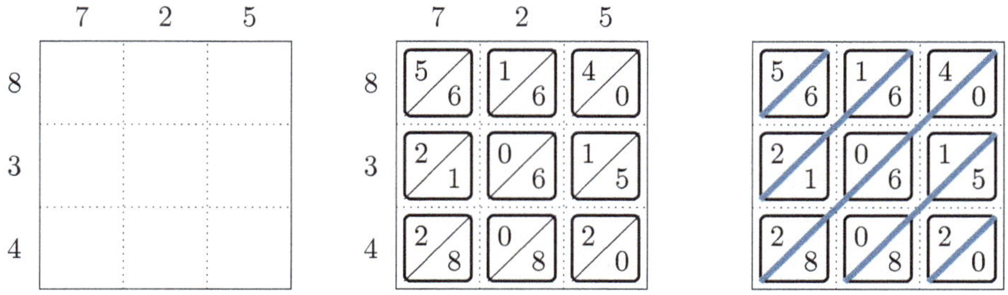

(a) (b) (c)

(c) We have thus obtained a rectangular array of 9 tiles, which form an assortment of 18 digits. The digits can be grouped into six "strips", separated by diagonal lines. The digits in each strip refer to a different decimal unit. The lower-right strip counts ones, the strip above it counts tens, the one above it counts hundreds, etc. Thus, the arrangement of the tiles represents the following assortment of decimal units:

0	ones
$5+2+8=15$	tens
$0+1+6+0+8=15$	hundreds
$4+6+0+1+2=13$	thousands
$1+6+2=9$	ten thousands
5	hundred thousands.

(d) What remains is to add up those decimal units. Alternatively, we may apply the standard addition algorithm (which in the present context might be called "diagonal addition"):

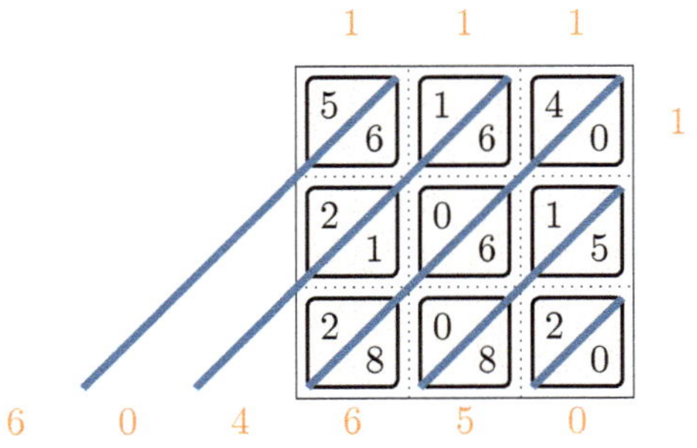

To conclude:

$$725 \times 834 = \boxed{604,650}$$

Did you see why this method works? To elucidate the idea behind Napier's method, try to find its connection with the expanded algorithm.

19.6.2 The binary algorithm

Consider the sequence of numbers starting with 1, with each subsequent number being twice its predecessor. These numbers are called **powers of two**, and they play a central role in computer science. The first ten numbers in this sequence are:

$$1, 2, 4, 8, 16, 32, 64, 128, 256, 512, 1024, \ldots.$$

A fundamental property of powers of two is that every natural number can be expressed in a unique way as a sum of non-repeating powers of two. For example, the unique representations of 73 and 171 as sums of non-repeating powers of two are:

$$73 = 64 + 8 + 1$$
$$171 = 128 + 32 + 8 + 2 + 1.$$

Given a natural number, we can decompose it into a sum of non-repeating powers of two is by a method called a **greedy algorithm**. Let's illustrate it for 171:

(a) Look for the greatest power of two less or equal to 171. It is 128.
(b) Find out the difference between 128 and 171. It is 43.
(c) Look for the greatest power of two less or equal to 43. It is 32.
(d) Find out the difference between 32 and 43. It is 11.
(e) Look for the greatest power of two less or equal to 11. It is 8.
(f) Find out the difference between 8 and 11. It is 3.
(g) Look for the greatest power of two less or equal to 3. It is 2.
(h) Find out the difference between 2 and 3. It is 1.
(i) Look for the greatest power of two less or equal to 1. It is 1.

(j) Find out the difference between 1 and 1. It is zero, hence we are done.
(k) It follows that $171 = 128 + 32 + 8 + 2 + 1$.

The binary multiplication algorithm is based, like all other multiplication algorithms, on the distributive property of multiplication. The only prerequisites are (i) knowing how to decompose a natural number into a sum of non-repeating powers of two, and (ii) the ability to multiply any number by 2.

Consider once again the product

$$73 \times 84.$$

Its evaluation using the binary algorithm proceeds as follows:

(a) Decompose the multiplier as a sum of non-repeating powers of two:

$$73 = 64 + 8 + 1.$$

(b) Apply the distributive property using that decomposition,

$$73 \times 84 = (64 \times 84) + (8 \times 84) + (1 \times 84).$$

(c) All the products on the right-hand side can be evaluated by multiplying 84 repeatedly by 2 (except for the third product which is trivial). Indeed,

$$\begin{aligned} 1 \times 84 &= 84 \\ 2 \times 84 &= 168 \\ 4 \times 84 &= 336 \\ 8 \times 84 &= 672 \\ 16 \times 84 &= 1344 \\ 32 \times 84 &= 2688 \\ 64 \times 84 &= 5376. \end{aligned}$$

(d) Adding up the three relevant products, we finally obtain:

$$5376 + 672 + 84 = \boxed{6132},$$

19.6.3 The peasants' algorithm

The last multiplication algorithm that we will present is called the **peasants' algorithm**. As its name indicates, it was widespread among uneducated populations (it was used in ancient Egypt). It is based on the exact same principles as the binary algorithm; only its implementation is different. The prerequisite skills are addition, multiplication by 2 and division by 2.

Consider once again the product

$$73 \times 84.$$

The peasants' algorithm proceeds as follows:

(a) Write down the two factors, one next to the other.
(b) Under one of the factors write the quotient of its division by 2. If this factor is odd, as is the case for 73, write only the integer part of the quotient—36.
(c) Under the other factor write the product of its multiplication by 2.
(d) Repeat this procedure until you obtain a quotient of 1.

÷2			×2
	73	84	
÷2	36	168	×2
÷2	18	336	×2
÷2	9	672	×2
÷2	4	1344	×2
÷2	2	2688	×2
÷2	1	5376	×2

(e) At this stage there are two columns of numbers: one column of decreasing numbers (left) and another column of increasing numbers (right). Proceed line by line: if the number on the left is even, cross out that line:

73	84
~~36~~	~~168~~
~~18~~	~~336~~
9	672
~~4~~	~~1344~~
~~2~~	~~2688~~
1	5376

(f) Add up the numbers on the right column that were not crossed out:

$$84 + 672 + 5376 = 6132.$$

Hocus pocus! This is the product.

Mathematical problems

Problem 19.1. Evaluate the following products:

$$40 \times 70$$
$$400 \times 70000$$
$$600 \times 500.$$

Problem 19.2. Evaluate the product

$$437 \times 39$$

using the following methods:

(a) The distributive property.
(b) The area model.
(c) The grid model.
(d) The expanded algorithm.
(e) The standard algorithm.
(f) Napier bones.
(g) The binary algorithm.
(h) The peasants' algorithm.

Problem 19.3. Evaluate the following products using mental calculations:

$$13 \times 7$$
$$32 \times 9$$
$$74 \times 5.$$

Problem 19.4. Investigate the relation between Napier's algorithm and the expanded algorithm.

Problem 19.5. Decompose the following numbers into non-repeating sums of powers of two:

$$314, \quad 2000, \quad 511, \quad \text{and} \quad 1024.$$

Problem 19.6. Investigate the peasants' algorithm: How is it related to the binary algorithm? In order to examine this question, evaluate the following products,

$$16 \times 3 \qquad \text{and} \qquad 15 \times 3,$$

using both methods.

Problem 19.7. Do we have a new commutative property?

(a) Verify that $24 \times 63 = 42 \times 36$.
(b) Verify that $93 \times 13 = 39 \times 31$.
(c) Is it always the case that the product of two two-digit numbers does not change if we interchange the order of the digits in both factors? If so, why? If not, under what conditions does this hold?

Problem 19.8. Place the digits 8, 5 and 2 in the following template, such that the product will be maximized:

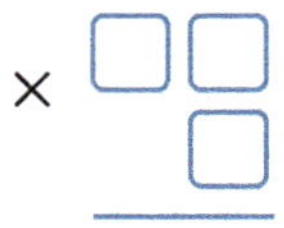

Repeat the same task using the digits 9, 6 and 5. Formulate a general rule for solving this type of problems (e.g., what should be the position of the largest digit?).

Problem 19.9. Place the digits 8, 5, 4 and 2 in the following template, such to maximize the product:

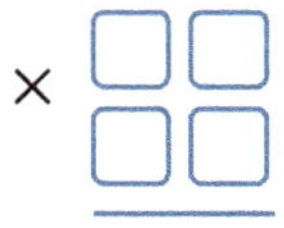

Repeat the same task using the digits 9, 8, 7 and 2. Formulate a general rule for solving this type of problems (e.g., What should be the position of the largest digit? What should be the position of the second largest digit? How many different solutions are there?).

Pedagogical problems

Problem 19.10. When Ms. Stryker checked Tal's homework on products of two-digit numbers, she discovered that Tal had made errors in all the exercises:

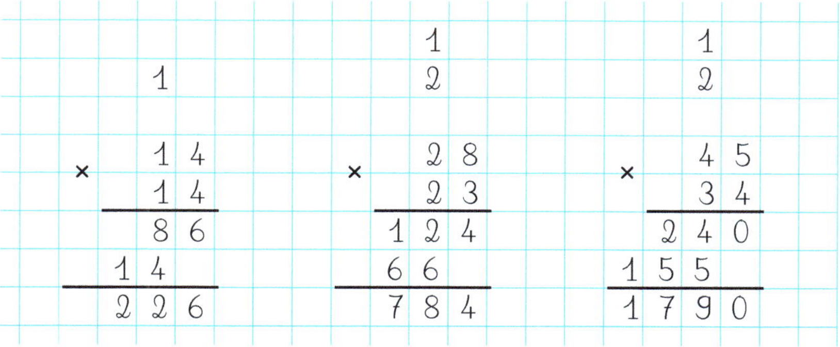

(a) Identify Tal's systematic error.
(b) How would you help Tal correct these errors?
(c) Think about additional systematic errors that children might do in exercises of this type.

Chapter 20

Division of Multi-Digit Numbers

This chapter is concerned with the evaluation of quotients of multi-digit numbers. It concludes a series of four chapters treating the evaluation of the four arithmetic operations with multi-digit operands (addition in Chapter 10, subtraction in Chapter 11 and multiplication in Chapter 19). This chapter only focuses on division without a remainder; division with a remainder will be dealt with in Chapter 22.

The standard division algorithm (also known as **long division**) is considered notoriously hard. Educators often raise doubts about whether proficiency in division should remain mandatory. Many feel that learning to divide multi-digit numbers is not worth the time, effort and frustration it raises among many children, compared to the alternative of using a calculator.

There are several good reasons to keep the division of multi-digit numbers in the curriculum :

(1) Evaluating arithmetic operations is the heart of the elementary school curriculum. Why make an exception for division? Proficiency in division is not beyond children's reach. Avoiding it sends a harmful message that there are topics which rely entirely on knowledge they possess, and yet are too difficult for them.
(2) It is through the endeavoring to divide large numbers that children sharpen their understanding of division and internalize its properties.
(3) Understanding division is the starting point for understanding fractions—the major hurdle in elementary school mathematics. Time and effort invested in deepening children's proficiency in division will prove rewarding at a later time.
(4) Introducing a challenging subject is also an opportunity to reveal the power of mathematics. In particular, breaking up a difficult task into several easier tasks is a central theme in mathematics.

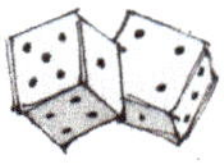

Activity: A first exposure to multi-digit division

As I advocated for the other three arithmetic operations, I recommend challenging children with evaluation problems prior to teaching them systematic methods of solution, thus encouraging them to develop a sense for numbers, as well as strategies of their own.

Consider, for example, the following question:

> 176 children attended a baseball tournament. They were divided into teams of 8 children each. How many teams participated in the tournament?

This question is likely to be hard for a child who hasn't learned how to evaluate division in this number range. However, many children may develop interesting ways to solve it, and the benefits of contemplating and discussing such personal strategies cannot be overrated.

Hurdles:

(a) Children struggle dividing large numbers while not mastering division within the multiplication table range.
(b) Children have a hard time memorizing the many stages of long division.
(c) Children adopt a how-to attitude; they get into the unhealthy habit of acting mechanically without looking for a meaning.
(d) Children don't have means to validate their solutions.

Remedies:

(a) Keep practicing both multiplication and division within the multiplication table range.
(b) Precede the use of structured algorithms with less structured methods, such as chunking.
(c) Demonstrate the division process using manipulatives emphasizing the decimal structure of multi-digit numbers; banknotes are excellent manipulatives for that purpose.
(d) Teach long division, demonstrating how each step in the algorithm manifests in a concrete model.
(e) Explore alternative methods of recording each step of the long division algorithm.
(f) Get children into the habit of validating quotients by performing the inverse multiplication operation.
(g) Get children into the habit of validating quotients using estimates; recall that estimates are sometimes as valuable as a precise answer.
(h) Encourage mental calculations; they may help internalizing the evaluation algorithms while building up self-esteem.

Opportunities:

Exploit division algorithms to reinforce in children's minds the principles underlying the decimal representation system.

20.1 Division without regrouping

We start by considering division problems in which the result— the quotient—can be evaluated, each decimal unit separately. In the next section, we will extend the scope to division problems requiring **regrouping**, namely, conversion between decimal units.

Consider the following arithmetic expression:

$$639 \div 3.$$

Division has two principal interpretations (Volume 1, p. 97), hence this quotient can be interpreted in two different ways:

(a) **Partitive division**: 639 items were divided equally among 3 sets. The quotient is the number of items in each set (i.e., the size of each **part**, or each **share**).

(b) **Quotative division**: 639 items were divided into sets containing 3 items each (i.e., **quotas** of 3 items). The quotient is the number of such sets.

In the following subsections we will examine several ways to evaluate this quotient.

20.1.1 Using the distributive property

Adopting the partitive division viewpoint, we are looking for the size of one share when 639 items are divided into 3 equal shares. We focus our attention on the decimal structure of the dividend, 639, and we reword the question in a way that emphasizes its structure as a composition of hundreds, tens and ones:

> 6 hundreds, 3 tens and 9 ones were divided into 3 equal shares. What is the size of each share?

Forget momentarily that hundreds, tens and ones are decimal units among which there exists a fixed conversion rate. If we replace them by three arbitrary nouns, such as *watermelons*, *apples*, and *cherries*, we obtain the following question:

> 6 watermelons, 3 apples and 9 cherries were divided into 3 equal shares. What is the size of each share?

What started as the evaluation of a quotient of a three-digit number has become a much simpler task: three independent quotient evaluations, all involving single-digit numbers. This question can be swiftly answered:

> The size of each share is 2 watermelons, 1 apple and 3 cherries.

Reverting watermelons to hundreds, apples to tens and cherries to ones:

The size of each share is 2 hundreds, 1 ten and 3 ones.

That is:

$$639 \div 3 = \boxed{213}.$$

This method of solution relies on the **distributive property** of division (Volume 1, p. 104). Written as an identity between arithmetic expressions, we decompose the dividend into a sum of hundreds, tens and ones and apply the distributive property,

$$639 \div 3 = (600 + 30 + 9) \div 3 = \underbrace{\underbrace{(600 \div 3)}_{200} + \underbrace{(30 \div 3)}_{10} + \underbrace{(9 \div 3)}_{3}}_{213}.$$

20.1.2 Algorithm

Arithmetic algorithms are conveniently carried out on grid paper, which facilitates both horizontal and vertical alignments of digits. The first step is writing down the operands. The two division operands are written horizontally, separated by a vertical bar, with the divisor at the left of the dividend (this differs from all other operations, in which the two operands are written one above the other). The space allocated to the (yet to be calculated) quotient is above the dividend, with like decimal units aligned vertically. Finally, a horizontal line is traced above the dividend, separating it from the space allocated to the quotient. The resulting arrangement is as follows:

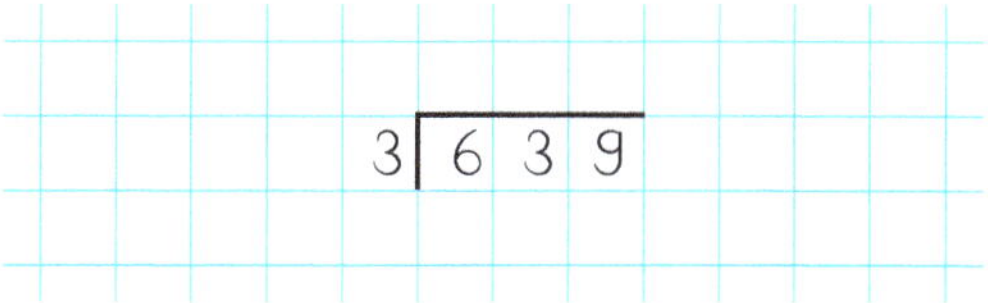

The quotient is evaluated digit after digit, from left to right–starting with the highest value digit. This order of proceeding differs from the other three operations, in which the evaluation always proceeds from right to left—starting with the ones digit.

The evaluation of $639 \div 3$ is greatly simplified by the fact that no regrouping is needed. Therefore, we present here a simplified version of the division algorithm, which is only valid in the absence of regrouping. Long division in its full glory is described in the next section.

The calculation proceeds as follows:

(a) **Divide hundreds**: 6 hundreds divided by 3 equal 2 hundreds. Write the digit 2 in the space allocated to the hundreds digit of the quotient.

(b) **Divide tens**: 3 tens divided by 3 equal 1 ten. Write the digit 1 in the space allocated to the tens digit of the quotient.

(c) **Divide ones**: 9 ones divided by 3 equal 3 ones. Write the digit 3 in the space allocated to the ones digit of the quotient.

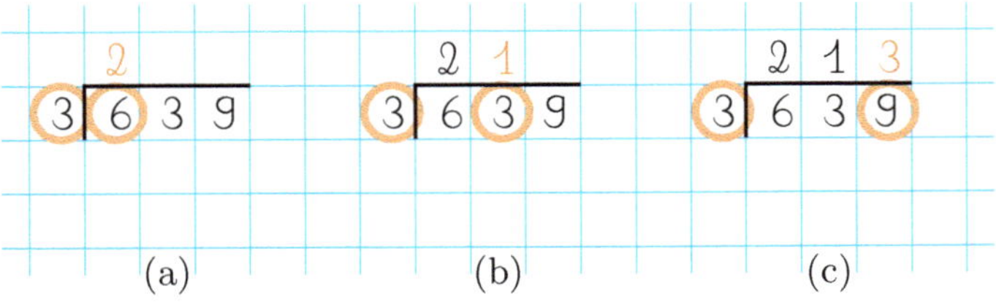

At the end of these steps, the quotient—213—is written above the dividend.

20.2 Division with regrouping

Next, we turn our attention to division problems that require regrouping of decimal units

20.2.1 First regroup and then evaluate

Consider the following arithmetic expression:

$$852 \div 3.$$

What is its value? As in the previous section, we rephrase the problem in a way that focuses the attention on the decimal structure of the dividend:

> 8 hundreds, 5 tens and 2 ones were divided into 3 equal shares. What is the size of each share?

Once again, lets replace *hundreds*, *tens* and *ones* by *watermelons*, *apples* and *cherries* and ask instead:

> 8 watermelons, 5 apples and 2 cherries were divided into 3 equal shares. What is the size of each share?

We run into a problem: how can we divide 8 watermelons into 3 equal shares without cutting up some of the watermelons? The same difficulty arises when it comes to dividing 5 apples and 2 cherries into three equal shares. It is here that we must exploit the fact that hundreds, tens and ones are not different fruits, but rather decimal units that unlike fruits, can be composed and decomposed. A ten can be decomposed into 10 ones, and a hundred can be decomposed into 10 tens.

The dividend, 852, represents a sum of hundreds, tens and ones, which we can display as follows:

8	5	2
H	T	O

In Chapters 10–12 we advocated the use of a non-standard decimal representation, whereby a decimal unit may repeat more than 9 times. Using this notation, we may represent that *same* number differently, by decomposing 2 hundreds (out of the given 8 hundreds) into 20 tens, yielding 6 hundreds, 25 tens and 2 ones. The resulting non-standard decimal representation of that number is:

6	25	2
H	T	O

Furthermore, we may decompose a ten into 10 ones, yielding 6 hundreds, 24 tens and 12 ones, which in non-standard representation reads:

6	24	12
H	T	O

Having changed the representation of the dividend (but not its value!), we rephrase the problem as follows:

> 6 hundreds, 24 tens and 12 ones were divided into 3 equal shares. What is the size of each share?

At last, we are facing three independent division problems, all within the multiplication table range. The answer is 2 hundreds, 8 tens and 4 ones, which together form a standard representation of a three-digit number. Thus, the solution to our problem is:

$$852 \div 3 = \boxed{284}.$$

This solution method, which is based on a judicious regrouping of the dividend, raises several concerns. How did we know which decimal units to decompose? Is such a strategy always possible? Can we expect third or fourth graders to be able to perform it? The truth is that the evaluation of quotients using this regrouping technique requires advanced skills. This is where long division comes in handy. It is a systematic algorithm for carrying out essentially the same procedure.

20.2.2 Illustrating the division process

Consider again the quotient:

$$852 \div 3.$$

Suppose that the objects quantified by the dividend were dollars (or any other currency). As advocated in Chapters 11 and 12, money is a useful manipulative, because the currency tallies with the decimal system: there are banknotes of $1, $10, and $100. In order to correctly simulate our decimal notation, we should not use other bills, like $5, $20 and $50 bills.

Using banknotes, our task is to divide 8 hundreds, 5 tens, and 2 ones into 3 equals sets, as depicted below:

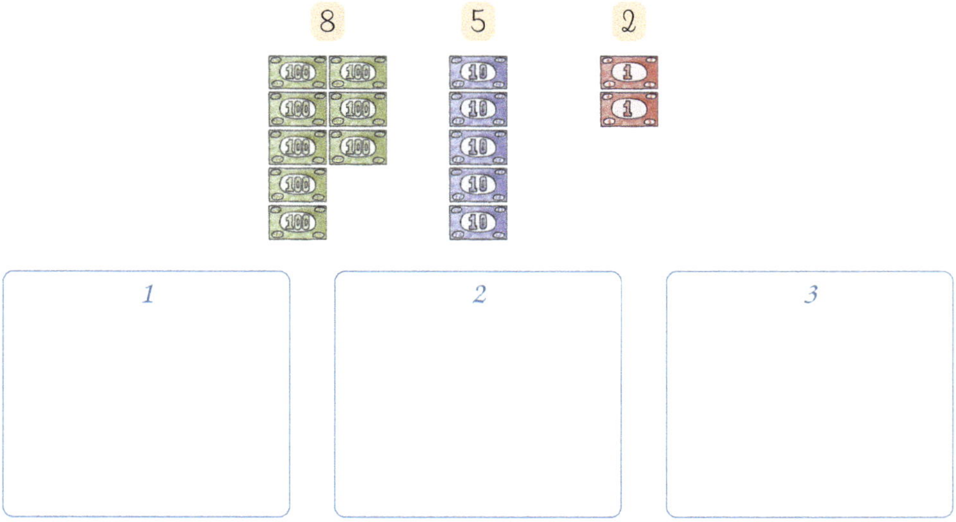

Division algorithms start with the leftmost digit, which in the present context refers to the $100 bills. We need to divide 8 hundred-dollar bills equally into 3 sets. The problem is that 8 is not divisible by 3. If we start distributing hundred-dollar bills equally among the three sets, we will end up placing 2 bills in each set, and have a **remainder** of 2 bills (the concept of a remainder will be discussed in detail in Chapter 22 below).

This is the first stage of the division algorithm: we distribute as many *whole* hundreds as possible—2 hundreds in each set—and retain the hundreds that were not distributed as a remainder, as depicted below:

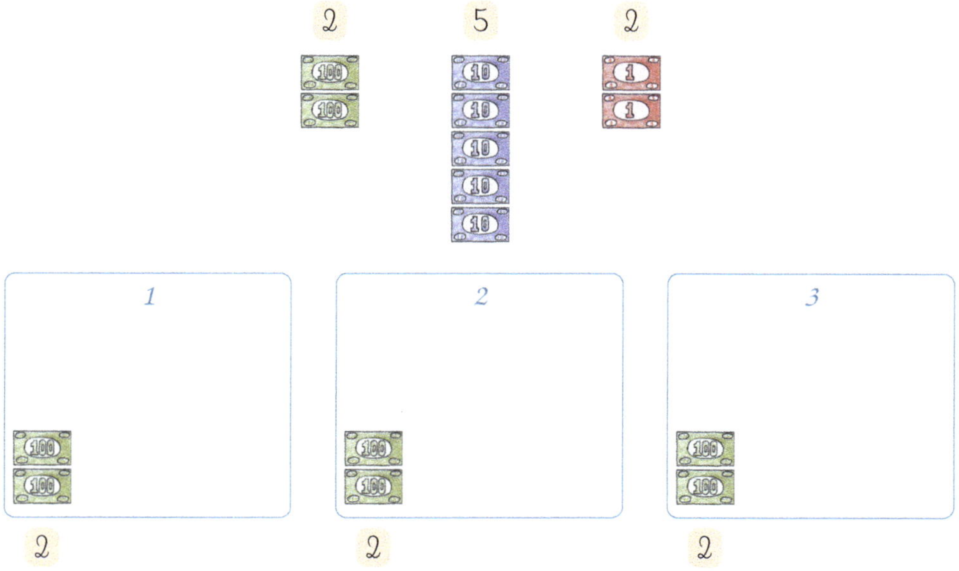

In the next stage, we exchange each of the remaining hundred-dollar bills into 10 ten-dollar bills. This adds 20 tens to the 5 given tens, yielding a total of 25 ten-dollar bills:

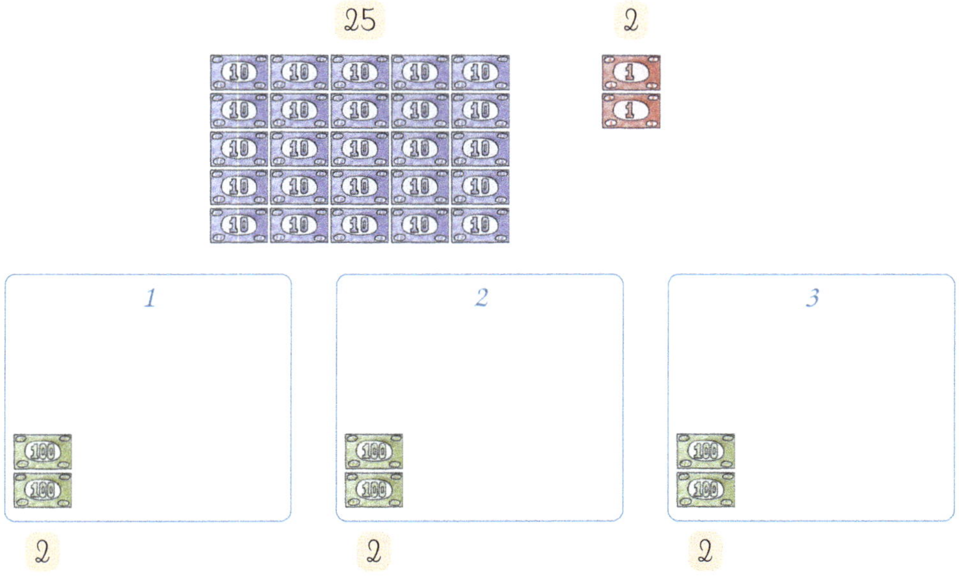

We proceed to distribute ten-dollar bills. We need to divide 25 ten-dollar bills equally into 3 sets. Once again, it is not possible to divide the entire quantity. If we start distributing ten-dollar bills equally among the three sets, we will end up placing 8 bills in each set, and have a remainder of 1 bill. Indeed, 8 is the largest number that when multiplied by 3 does not exceed 25. The result will be as follows:

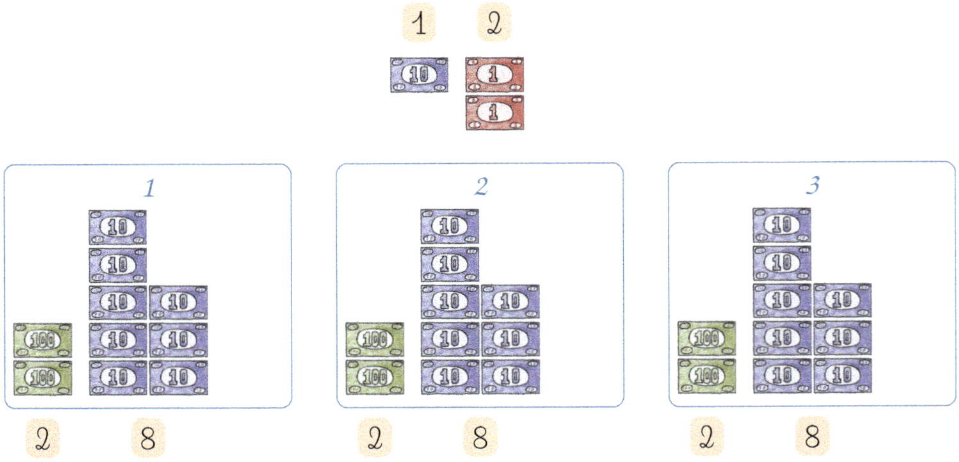

We now exchange the remaining ten-dollar bill into 10 one-dollar bills. This adds 10 ones to the 2 ones that were at the outset, yielding a total of 12 one-dollar bills:

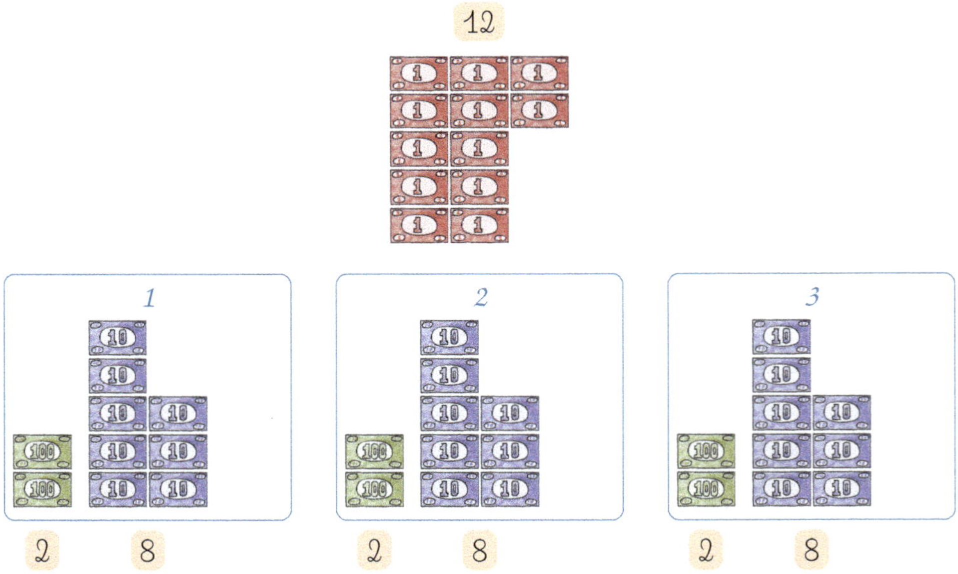

We conclude the procedure by dividing the one-dollar bills. We divide 12 one-dollar bills equally among the three sets, 4 bills in each set, this time without any remainder:

We have thus divided the \$852 equally between the three sets, each set comprising 2 hundreds, 8 tens and 4 ones; that is to say,

$$852 \div 3 = \boxed{284}.$$

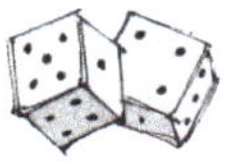

Activity: Practicing division using manipulatives

Division with manipulatives, such as banknotes, is a valuable way to understand and assimilate the division algorithm. I strongly recommend solving numerous division problems in concrete settings. Provide a detailed verbal description of each stage (e.g., "we need to divide 8 hundreds evenly among 3 sets; however, we can divide at most 6 of them," or "if we exchange 2 hundreds into tens, we get 20 tens that add to the 5 tens we already have").

It is worthwhile to raise the difficulty level gradually. For example:

(1) $848 \div 4$: No regrouping is needed.
(2) $483 \div 3$: Requires to regroup once.
(3) $714 \div 6$: Requires to regroup twice.
(4) $873 \div 9$: The quotient has fewer digits than the dividend.

20.2.3 Long division: Step-by-step

We next demonstrate how to evaluate the quotient

$$852 \div 3$$

using the standard division algorithm. While doing so, it is instructive to relate each stage to its concrete counterpart when using banknotes. Having in mind a concrete image helps understanding the algorithm and memorizing it.

As in the previous section, we start by writing down the two operands. We then proceed as follows:

(a) **Divide hundreds**: We have to divide 8 hundreds by 3. Since 8 is not divisible by 3, we are looking, in fact, for the largest number, which when multiplied by 3, yields a result not exceeding 8. This number is 2. We write 2 in the space allocated for the hundreds digit of the quotient, above the hundreds digit of the dividend.

(b) **Find how many hundreds were divided**: We were supposed to divide 8 hundreds, however, we only divided as many hundreds as we could without decomposing hundreds into smaller decimal units. To find how many hundreds were actually divided without having to memorize information from previous stages, we multiply the 2 hundreds in the quotient by the divisor, 3. The product is 6 hundreds. We write 6, representing the hundreds that were actually divided, under the 8 representing the hundreds that ought to be divided.

(c) **Calculate the remainder**: To find how many hundreds have not yet been divided, we subtract the 6 hundreds that were actually divided from the 8 hundreds that ought to be divided. We obtain a remainder of 2 hundreds, which we write under both numbers. Note that this digit, which represents hundreds, is aligned with the hundreds digit of the dividend.

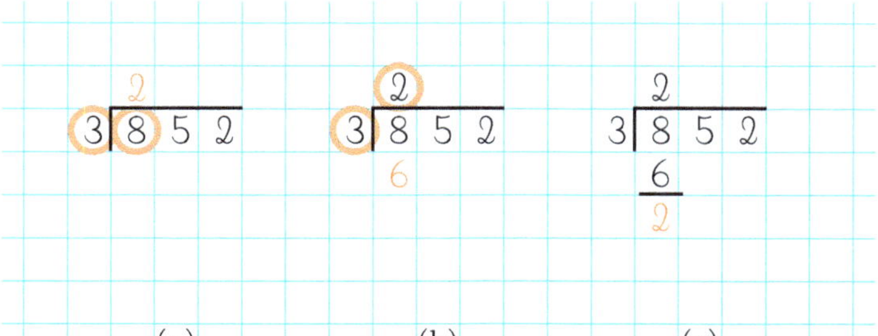

(a) (b) (c)

(d) **Decompose hundreds into tens**: We decompose the remaining 2 hundreds into 20 tens. In practice, we "pull down" the tens digit of the dividend, 5, to the right of the number of remaining hundreds. We obtain a two-digit number, 25, representing the number of tens that have yet to be divided.

(e) **Divide tens**: We have to divide 25 tens by 3. Since 25 is not divisible by 3, we are looking, in fact, for the largest number, which when multiplied by 3, yields a result not exceeding 25. This number is 8. We write 8 in the space allocated for the tens digit of the quotient.

(f) **Find how many tens were divided**: We were supposed to divide 25 tens, however, we only divided as many tens as we could without decomposing them into smaller decimal units. To find how many tens were actually divided, we multiply the 8 tens in the quotient by the divisor, 3. The product is 24 tens. We write 24, representing the tens that were actually divided, under the 25 representing the tens that ought to be divided.

(g) **Calculate the remainder**: To find how many tens have not yet been divided, we subtract the 24 tens that were divided from the 25 tens that ought to be divided. We obtain a remainder of 1 ten, which we write under both numbers. Note that this number, which represents tens, is aligned with the tens digit of the dividend.

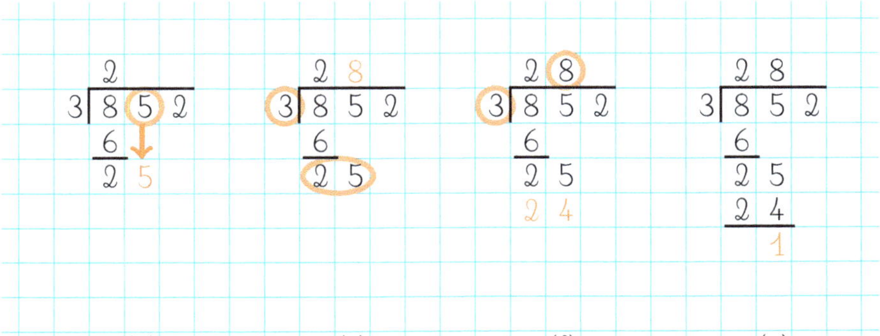

(d) (e) (f) (g)

(h) **Decompose tens into ones**: We decompose the remaining ten into 10 ones. In practice, we "pull down" the ones digit of the dividend, 2, to the right of the number of remaining tens. We obtain a two-digit number, 12, representing the number of ones that have yet to be divided.

(i) **Divide ones**: We have to divide 12 ones by 3. Once again, we are looking for the largest number, which when multiplied by 3, yields a result not exceeding 12. This number is 4. We write the digit 4 in the space allocated for the ones digit of the quotient.

(j) **Find how many ones were divided**: We were supposed to divide 12 ones. To find how many ones were actually divided, we multiply the 4 ones in the quotient by the divisor, 3. The product is 12 ones. We write 12, representing the ones that were actually divided, under the 12 representing the ones that ought to be divided.

(k) **Calculate the remainder**: To find how many ones have not yet been divided, we subtract the 12 ones that were actually divided from the 12 ones that ought to be divided. The difference in this case is zero, meaning that there is no remainder. This indicates that we completed the division process.

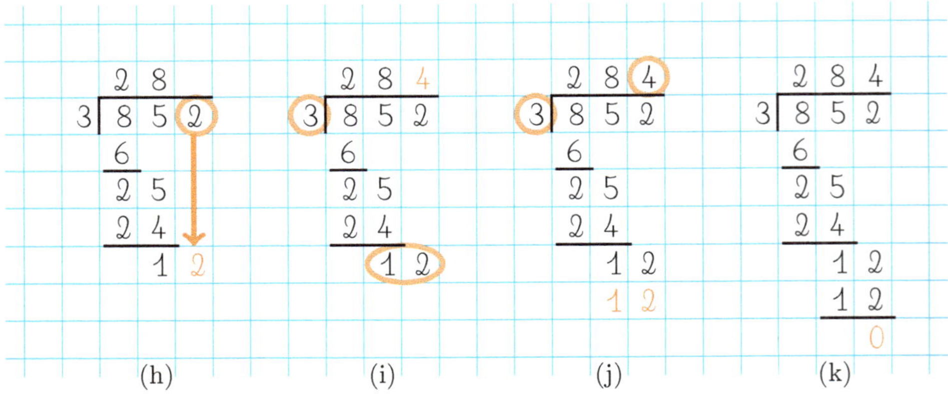

Please note: an algorithm is not a substitute for common sense. It is strongly recommended to examine the plausibility of a solution at the end of every calculation, especially after long calculations which are more prone to errors. For example, one could check whether the solution 284 is plausible, by verifying that 200×3 is less than 854, whereas 300×3 is greater than 852.

20.3 Division by a multi-digit number

Thus far, we have dealt with quotients in which the divisor is a single-digit number. The same division algorithm applies to multi-digit divisors as well. The level of difficulty, however, is higher, as each step presents us with a division problem beyond the multiplication table range.

Consider the following quotient:

$$323 \div 17.$$

We might be tempted to evaluate it by distributing the divisor into $17 = 10 + 7$. But remember, division is only distributive with respect to the dividend; the divisor does not satisfy a distributive property (Volume 1, p. 105).

To evaluate this quotient, we follow the long division algorithm step-by-step:

(a)–(c) **Divide hundreds**: We divide 3 hundreds by 17. Clearly, no *whole* hundreds are divided, hence the hundreds digit of the quotient is zero.

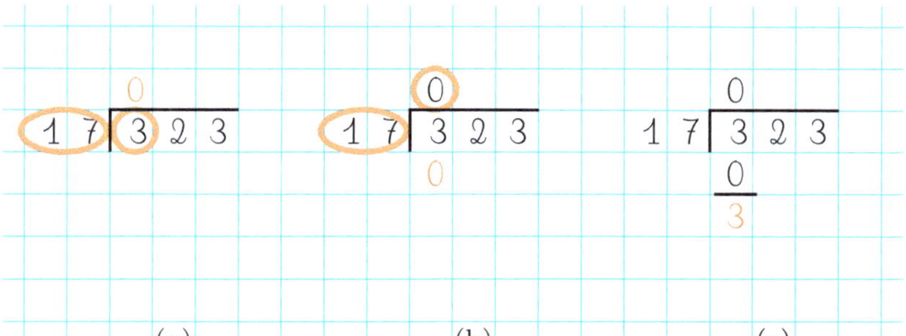

(a) (b) (c)

(d)–(g) **Divide tens**: We decompose the 3 hundreds into 30 tens, which together with the 2 tens that we had at the outset make 32 tens. We divide the 32 tens by 17, giving a quotient of 1 and a remainder of 15:

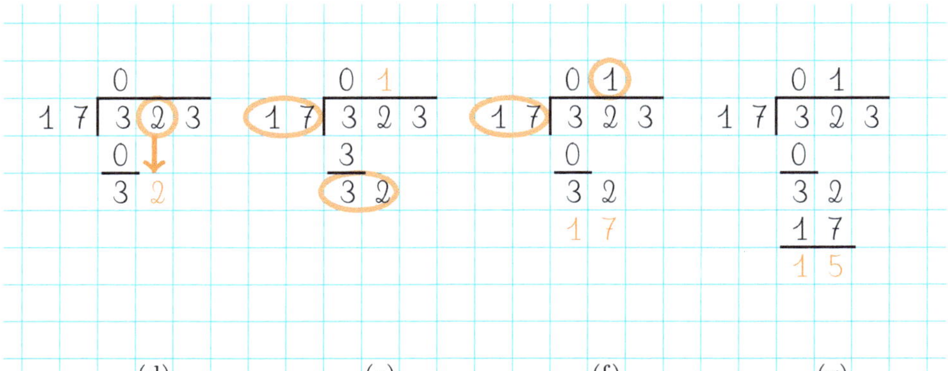

(d) (e) (f) (g)

(h)–(k) **Divide ones**: We decompose the 15 tens into 150 ones, which together with the 3 ones that we had at the outset make 153 ones. We divide the 153 ones by 17, giving a quotient of 9 and no remainder. We conclude that

$$323 \div 17 = \boxed{19}.$$

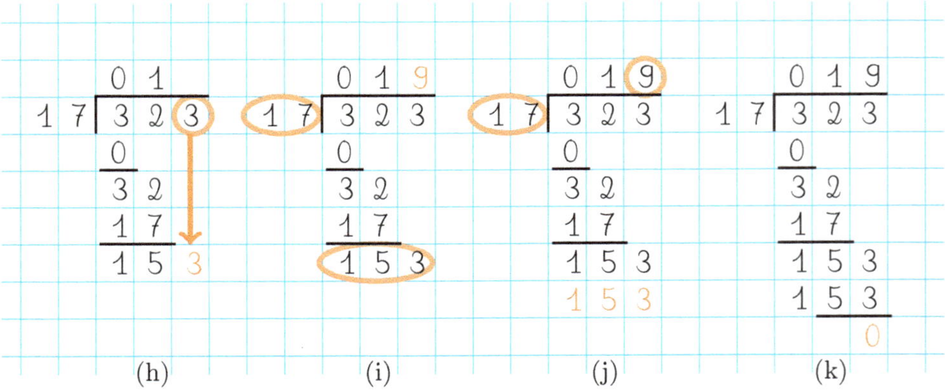

Please note: it is always recommended to check the results of a division operation through the inverse multiplication problem. In this case:

$$17 \times 19 = \boxed{323}.$$

Comments

(1) Chapter 25, deals with **division signs**, which are quick and efficient ways of identifying whether a certain number is divisible by another.

(2) The major difficulty in the division algorithm is that at each stage, we need to find the largest factor, which when multiplied by the divisor, yields a product not exceeding the number being divided. Finding this largest factor requires some kind of "backward thinking", which is a non-trivial skill. When observing children carry out long division, we very often find alongside the solution scribbles of multiplications reflecting their attempts to determine that largest factor.

(3) An additional difficulty with the division algorithm is the profusion of steps (division, multiplication, subtraction, regrouping), along with the complicated way of recording all the various intermediary results. The mathematical challenges are thus augmented by organizational challenges. Many teachers have attempted to find solutions to these difficulties.

We next present an alternative division method, which alleviates some of the notational complications. I have heard more than one teacher say that children find this notation way easier than the standard algorithm.

(a) Write the arithmetic expression, leaving some space between the digits:

$$8\ 5\ 2 \div 3.$$

(b) **Divide hundreds**: 8 hundreds divided by 3 yield a quotient of 2 hundreds and a remainder of 2 hundreds. Write the quotient, 2, on the right-hand side, and write the remainder, 2, as a subscript at the left of the tens digit of the dividend, turning it into 25 tens:

$$8\ {}_{2}5\ 2 \div 3 = 2.$$

(c) **Divide tens**: 25 tens divided by 3 yield a quotient of 8 tens and a remainder of 1 ten. Write the quotient, 8, on the right-hand side, and write the remainder, 1, as a subscript at the left of the ones digit of the dividend, turning it into 12 ones:

$$8\ {}_{2}5\ {}_{1}2 \div 3 = 2\ 8.$$

(d) **Divide ones**: 12 tens divided by 3 yield a quotient of 4 ones without a remainder. Write the quotient, 4, on the right-hand side:

$$8\ {}_{2}5\ {}_{1}2 \div 3 = 2\ 8\ 4.$$

I don't think that this way of evaluating division resolves all problems (in particular, some steps in the calculation are not written down). It is nevertheless recommended to experiment with various evaluation algorithms if your experience indicates that they make it easier for some of the children.

20.4 Chunking

A disadvantage of algorithms such as long division is that children's intuition and number sense is often lost within the long sequence of steps. An alternative to the standard division algorithm, which is gaining popularity in many countries, utilizes a less structured strategy. The idea, which is also based on the distributive property, is to decompose the dividend into "chunks", whence the name given to this method—**chunking**.

We will demonstrate the chunking method by evaluating the following expression:

$$455 \div 13.$$

Adopting a quotative division approach, we need to find how many times 13 is contained in 455. Seemingly, it is hard to answer this question without a systematic algorithm, but a possible alternative is to attack the problem in stages.

(a) Clearly, 13 is contained in 455 more than 10 times because $10 \times 13 = 130$, which is far less than 455. We remove 130 from 455, leaving 325, keeping in mind that we have removed a chunk of 10 times 13 from the dividend.

(b) Similarly, 13 is contained in 325 more than 10 times. We remove 130 from 325, leaving 195, keeping in mind that we have removed an additional chunk of 10 times 13.

(c) Once again, 13 is contained in 195 more than 10 times. We remove 130 from 195, leaving 65, keeping in mind that we have removed an additional chunk of 10 times 13.

(d) 13 is contained in 65 more than 2 times, because $2 \times 13 = 26$, which is less than 65. We remove 26 from 65, leaving 39, keeping in mind that we have removed an additional chunk of 2 times 13.

(e) 13 is contained in 39 exactly 3 times.

(f) We add up the various chunks: the number of times that 13 is contained within 455 is

$$10 + 10 + 10 + 2 + 3 = 35,$$

namely,

$$455 \div 13 = \boxed{35}.$$

Please note: the chunking method is *not* an algorithm. There is no pre-determined order of proceeding. Different people will decompose the same dividend into different chunks. Educators claim that the lack of order is in fact a major advantage of the chunking method, as it hinges on a sense for numbers, rather than on a sequence of mechanical operations.

Mathematical problems

Problem 20.1. Evaluate the following quotients using the long division algorithm, the alternative procedure described on p. 45 and the chunking method:

$$693 \div 3$$
$$564 \div 6$$
$$407 \div 11.$$

Problem 20.2. In Section 20.2.1 we evaluated the quotient $852 \div 3$ using a "first regroup and then evaluate" approach. The idea behind this method is the following:

- The number of hundreds, 8 is not a multiple of 3. The largest multiple of 3 not exceeding 8 is 6. The difference is 2 hundreds, which we decompose into tens. In non-standard decimal representation, we obtain

6	25	2
H	T	O

- The number of tens, 25 is not a multiple of 3. The largest multiple of 3 not exceeding 25 is 24. The difference is 1 ten, which we decompose into ones. In non-standard decimal representation, we obtain

6	24	12
H	T	O

- Now all three "digits" are divisible by 3 and we can carry out the division digit-by-digit.

(a) Write down the number 714 in non-standard decimal notation so that it can be divided by 3 digit-by-digit.
(b) Write down the number 714 in non-standard decimal notation so that it can be divided by 6 digit-by-digit.
(c) Write down the number 714 in non-standard decimal notation so that it can be divided by 7 digit-by-digit.

Problem 20.3. Evaluate the quotient

$$834 \div 3$$

starting from the ones digit, proceeding from right to left. Explain the disadvantages of this approach.

Pedagogical problems

Problem 20.4. Evaluate the three quotients presented in Problem 20.1. This time, illustrate the process with banknotes (for instance, Monopoly money). After each step, write down its counterpart in the long division algorithm.

Chapter 21

The Order of Operations

The four operations of arithmetic—addition, subtraction, multiplication and division—are ***binary operations***: they operate on *two* numbers called **operands**, and produce a single number as output. Yet, several arithmetic operations can be combined together to form compound arithmetic expressions. In such case, it is unclear in what order should operations be performed. As we will see, different orders of operarions yield different values. Thus, a convention is needed in order to read each compound arithmetic expression unambiguously.

Hurdles:

(a) Children evaluate compound arithmetic expressions incorrectly.
(b) Children don't understand the rationale behind the evaluation rules.

Remedies:

(a) Keep practicing the evaluation of compound arithmetic expressions.
(b) Emphasize the fact that the order of operations is nothing more than a convention.

Opportunities:

Good command of the order of operations is a prerequisite for algebraic proficiency in middle school.

21.1 Problem statement

In Chapter 4 we encountered arithmetic expressions involving the addition of three or more addends, say,

$$12 + 6 + 7.$$

A priori, this expression is ambiguous. It is not clear whether we first have to add 12 and 6, and then add 7 to that sum, or conversely, add to 12 the sum of 6 and 7. Addition, as we learned, satisfies an **associative property**, meaning that the value of this expression is the same in either way (Volume 1, p. 36):

$$\underbrace{\underbrace{12+6}_{18}+7}_{25} = \underbrace{12+\underbrace{6+7}_{13}}_{25}.$$

In contrast, consider the following expression,

$$5 + 7 \times 3,$$

which combines addition and multiplication. If we first add 5 and 7 and then multiply the sum by 3, we obtain 36:

$$\underbrace{\underbrace{5+7}_{12}\times 3}_{36}.$$

If, on the other hand, we add to 5 the product of 7 and 3, we obtain 26:

$$\underbrace{5+\underbrace{7\times 3}_{21}}_{26}.$$

That is, a sequence of arithmetic operations combining addition and multiplication does not satisfy an associative property. The order of the operations affects the value of the expression.

This is not our first encounter with non-associativity. Both subtraction and division are not associative (Volume 1, pp. 58,103). In the event of two consecutive subtractions or two consecutive divisions, the convention is that the operations are performed in their order of appearance, from left to right. The question is whether operations are always performed sequentially, even when, for instance, addition is followed by multiplication. And in our case, what is the value of the arithmetic expression $5 + 7 \times 3$? 36 or 26?

In Chapter 3, we visualized the arithmetic operations as machines with two input ports and one output port. When two numbers are entered into the machine's input ports, the machine computes a number and returns it as its output. A dedicated machine corresponds to each operation: there are addition machines, subtraction machines, multiplication machines and division machines.

The two ways in which we could interpret the arithmetic expression $5 + 7 \times 3$ can be illustrated with addition and multiplication machines. The case where addition precedes multiplication is illustrated below:

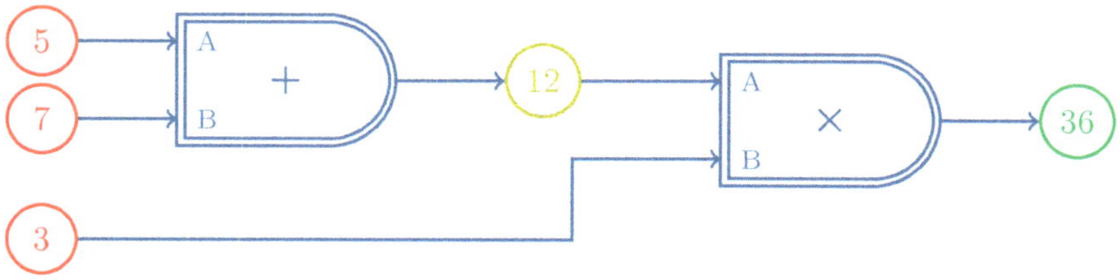

This is an example of a **computation network** (the term "network" has the same meaning as in "electric network" or as in "social network"). In this illustration and in those to follow, the inputs—namely, the numbers appearing in the arithmetic expression—are marked in red, intermediate evaluations are marked in yellow, and the output—namely, the value of the arithmetic expression—is marked in green.

The computation network corresponding to the case where multiplication precedes addition is presented below:

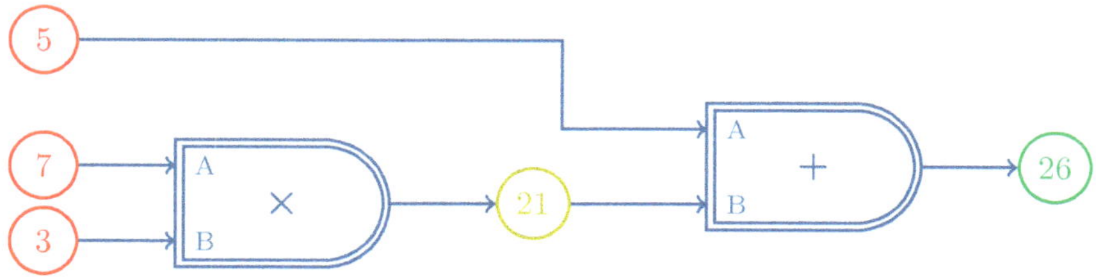

Comments

(1) The conventions determining the order of the arithmetic operations were established during the 17th century along with the development of algebra.

(2) The determination of the order of operations is essentially not a mathematical issue; it is more of a grammatical issue, deriving from the need to prevent ambiguity. One has to know these conventions in order to evaluate arithmetic expressions correctly. Yet, keep in mind that conventions are arbitrary.

(3) Ambiguous expressions occur not just in mathematics, but also in everyday spoken language. Consider the following sentence: "I see a man on a hill with a telescope". Does it mean that there's a man on a hill, and I'm watching him with my telescope? Does it mean that there's a man on a hill, whom I'm seeing, and he has a telescope? Does it mean there's a man, and he's on a hill that also has a telescope on it? Without proper punctuation marks, it is impossible to parse the sentence and determine the correct meaning. Punctuation marks help us determine the correct interpretation in spoken language, just as parentheses

and other conventions help us determine the correct interpretation of arithmetic expressions.

21.2 From computation networks to arithmetic expressions

Consider the following computation network:

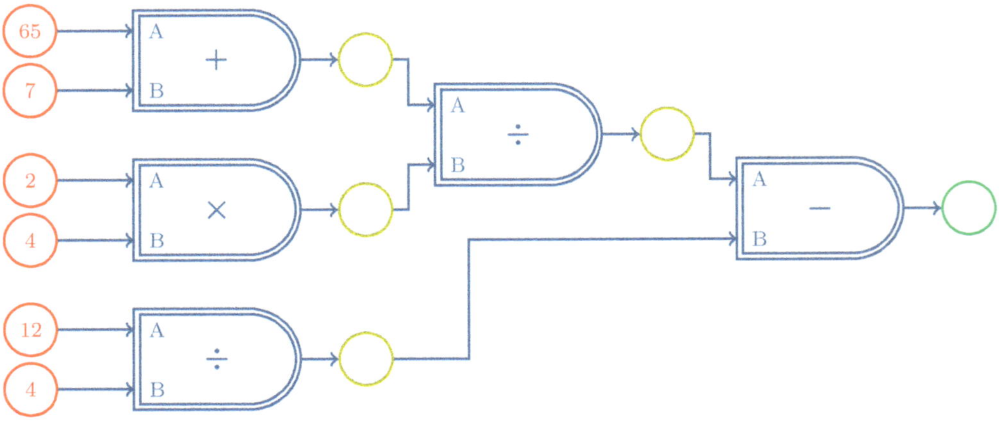

This computation network comprises five machines. It takes six inputs, and produces four intermediate evaluations and one final output. (Is it a coincidence that the total number of machines is one less than the number of inputs?)

Is there any ambiguity in how to evaluate the output of this computation network? No.

The order of operations is dictated by the structure of the network. A machine cannot operate before the machines that precede it have generated their output.

The output of this computation network is depicted below:

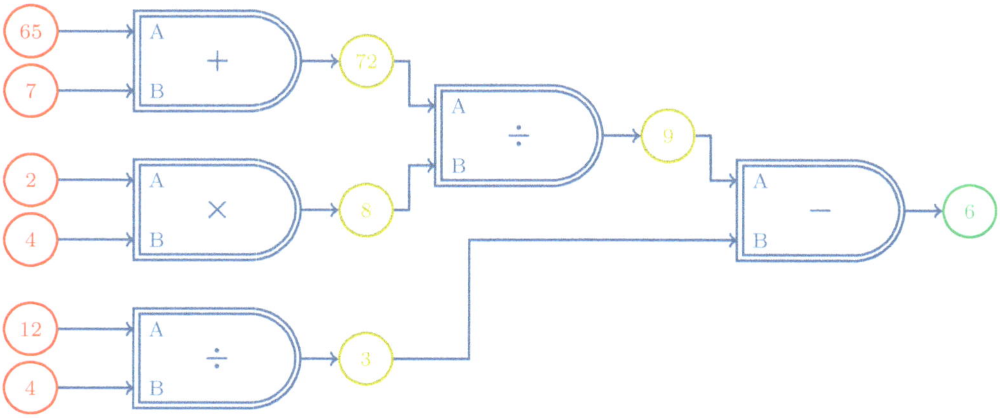

Unlike computation networks that dictate the order of operations, arithmetic expressions may be ambiguous. Therefore, we have to augment arithmetic expressions with some "network structure" guaranteeing that each expression can be interpreted in only one way. By convention, this structure uses two devices: (1) **parentheses** and (2) conventions **prioritizing** multiplication and division over addition and subtraction.

21.3 Parentheses

Parentheses are used in arithmetic expressions to set the precedence in the order of evaluation. When part of an arithmetic expression is enclosed in parentheses, it must be evaluated before it can be used as an operand for operations on either of its sides.

Consider for example the arithmetic expression

$$(5 + 7) \times 3,$$

which exhibits one pair of parentheses enclosing the expression $5 + 7$. In this case, the multiplication operation at the right of the parentheses cannot be evaluated prior to the expression within the parentheses. We first evaluate $5 + 7$, yielding 12, and then multiply this sum by 3, yielding 36.

In contrast, consider the arithmetic expression

$$5 + (7 \times 3).$$

In this case, the addition operation at the left of the parentheses cannot be evaluated prior to the expression within the parentheses. We first evaluate 7×3, yielding 21, and then add this product to 5, yielding 26.

Note the analogy between the structure dictated by parentheses and the structure dictated by computation networks. Below we will see that a systematic use of parentheses creates a one-to-one correspondence between computation networks and arithmetic expressions.

Examine the computation network on p. 52 once again, this time from the end to the beginning. The final output is generated by a subtraction machine; hence, we can represent it schematically as follows:

$$\text{output} = \text{minuend} - \text{subtrahend}.$$

The minuend is the output of an "upstream" division machine, which can be written as (dividend ÷ divisor). We enclose the expression in parentheses since it must be evaluated prior to serving as an operand for the subtraction operation. We marked the parentheses in red to differentiate them from other parentheses that will appear further on.

The subtrahend is also an output of a division machine; hence, we will write it as (dividend ÷ divisor). Putting everything so far together, we obtain

$$\text{output} = (\text{dividend} \div \text{divisor}) - (\text{dividend} \div \text{divisor}).$$

Next consider the expression within the red parentheses. The dividend in that expression is the output of an addition machine, which we write as (addend + addend), whereas the divisor in that same expression is the output of a multiplication machine, which we write as (multiplier × multiplicand). Putting it together, we obtain

$$\text{expression} = ((\text{addend} + \text{addend}) \div (\text{multiplier} \times \text{multiplicand})) - (\text{dividend} \div \text{divisor}).$$

This expression comprises 6 operands, which are the inputs to the network. Substituting with numbers, we finally obtain the arithmetic expression,

$$\text{output} = ((65 + 7) \div (2 \times 4)) - (12 \div 4).$$

This expression, which uses four pairs of parentheses, has the same value as the output of the computation network on p. 52. Like the computation network, its evaluation is unambiguous. The left-most division operation and the subtraction operation cannot be evaluated before the expressions enclosed within the inner parentheses. The parentheses impose that we have to evaluate first the expressions:

$$(65 + 7) = 72$$
$$(2 \times 4) = 8$$
$$(12 \div 4) = 3.$$

Substituting the result of these intermediate evaluations in the original arithmetic expressions we get

$$\text{output} = (72 \div 8) - 3.$$

Following the same logic, we note that the subtraction cannot be evaluated until the expression within the parentheses to its left is evaluated,

$$(72 \div 8) = 9.$$

Substituting, we finally obtain

$$\text{expression} = 9 - 3 = 6.$$

We can also display the order of operations in one line:

$$\underbrace{\underbrace{\big(\underbrace{(65+7)}_{72} \div \underbrace{(2\times 4)}_{8}\big)}_{9} - \underbrace{(12\div 4)}_{3}}_{6}.$$

We have thus shown how a computation network can be converted into an unambiguous arithmetic expression, with parentheses playing the role of punctuation marks.

What more is there to say? Mathematics aims for succinctness. Moreover, expressions comprising multiple pairs of parentheses are hard to decipher (which is the reason we used different colors to differentiate them). Augmenting the conventions with rules of precedence among operations makes it possible to reduce the number of parentheses without compromising clarity, and without introducing ambiguity.

21.4 Precedence of operations

Once again, consider the arithmetic expression,

$$((65+7) \div (2\times 4)) - (12 \div 4).$$

The reason this expression can only be evaluated in one way is that each mathematical operation, along with its two operands, is enclosed in parentheses. As a result, each of the six numbers can be associated with only one operation. For example, the number 7 is an operand of the addition operation to its left. The presence of a parenthesis to its right precludes its association to the adjacent division operation. Likewise, the number 12 can only be associated with the division operation to its right. The nesting of parentheses within parentheses guarantees that as we evaluate an operation within an inner pair of parentheses, we keep generating expressions that can be evaluated unambiguously.

In contrast, when considering the arithmetic expression

$$3 + 6 \times 8,$$

it is unclear whether 6 is an operand of the addition operation to its left or of the multiplication operation to its right. As we have seen, the two possibilities produce different results.

To remove all ambiguity, while keeping the number of parentheses to a minimum, the following four conventions were determined:

(a) Expressions within parentheses are evaluated first. In the case of nested parentheses, the inner ones are evaluated first.

(b) Multiplication and division precede addition and subtraction.
(c) Multiplication and division are carried out from left to right, by order of appearance.
(d) Addition and subtraction are carried out from left to right, by order of appearance.

Let's put those rules in action by evaluating the arithmetic expression

$$53 + (8 \div 2) - 5 \times 8 \div 4 - 4 \times 7.$$

(1) Evaluate the expression within the parentheses:

$$53 + \underbrace{(8 \div 2)}_{4} - 5 \times 8 \div 4 - 4 \times 7.$$

(2) Evaluate the two multiplications and the division by their order of appearance, from left to right:

$$53 + 4 - \underbrace{5 \times 8}_{40} \div 4 - 4 \times 7,$$

$$53 + 4 - \underbrace{40 \div 4}_{10} - 4 \times 7,$$

and

$$53 + 4 - 10 - \underbrace{4 \times 7}_{28}.$$

(3) Evaluate the addition and the two subtractions operations by their order of appearance, from left to right:

$$\underbrace{53 + 4}_{57} - 10 - 28.$$

$$\underbrace{57 - 10}_{47} - 28.$$

and

$$47 - 28 = 19.$$

Yet, keep in mind: instead of adopting complex conventions we could have only relied on parentheses:

$$((53 + (8 \div 2)) - ((5 \times 8) \div 4)) - (4 \times 7).$$

Note the six pair of parentheses, compared to only one pair of parentheses in the original expression. This is precisely the reason why precedence rules were added.

21.4.1 Deleting redundant parentheses

Having enumerated the four precedence rules, let us reconsider the arithmetic expression

$$((65 + 7) \div (2 \times 4)) - (12 \div 4).$$

This expression has four pairs of parentheses. Let's see if some of them can be deleted without affecting the order of operations.

(a) Examine the red parentheses: They enclose a division problem (the facts that both dividend and divisor are themselves arithmetic expressions is irrelevant to the present analysis). To their right is a subtraction. Whether we retain the parentheses or delete them, division precedes subtraction. Therefore, we can delete these parentheses without affecting the order of operations:

$$(65 + 7) \div (2 \times 4) - (12 \div 4).$$

(b) Examine the green parentheses: They enclose an addition problem, whereas to their right is a division problem. If we retain the parentheses then addition is carried out first. If we delete them, then division is carried out first. Therefore, we cannot delete these parentheses without affecting the order of operations.

(c) Examine the yellow parentheses: They enclose a multiplication problem. To their left is a division problem, and to their right is a multiplication problem. If we retain the parentheses then multiplication precedes division; however, if we delete them, then division precedes multiplication. Therefore, we cannot delete these parentheses without affecting the order of operations.

(d) Finally, examine the blue parentheses: They enclose a division problem, whereas to their left is a subtraction problem. Whether we retain the parentheses or delete them, division precedes subtraction. Therefore, we can delete these parentheses without affecting the order of operations:

$$(65 + 7) \div (2 \times 4) - 12 \div 4.$$

This last expression is the most succinct way to express the computation network on p. 52.

Activity: Digits and Letters

The French TV program, *des chiffres et des lettres* (literally, digits and letters) is a game that challenges computation and language capabilities of two contestants. The program was launched in France in 1965, and is the oldest French TV program of our time.

In the numerical part of the game, the contestants are presented a three-digit number, along with six randomly chosen numbers drawn from the following list,

$$1 \quad 2 \quad 3 \quad 4 \quad 5 \quad 6 \quad 7 \quad 8 \quad 9 \quad 10 \quad 25 \quad 50 \quad 75 \quad 100.$$

Each of these numbers can appear more than once.

The contestants have 45 seconds to combine those six numbers using the four arithmetic operations, in order to reach the three-digit target number. Each number, as well as each intermediate result, can be used at most once. Suppose, for example, that a contestant multiplied 25×4 getting 100. The numbers 25 and 4 can no longer be used, but 100 can be used further on in the calculation. It is not mandatory to use all six numbers or all four operations. The winner is the one who comes closest to the target number.

Let's consider an example: suppose that the six numbers are

8 2 4 6 8 9

and the target number is

594.

As it turns out, the target number can be obtained as follows:

$$8 + 8 = 16$$

$$16 \times 4 = 64$$

$$64 + 2 = 66$$

$$66 \times 9 = 594.$$

As earlier in this chapter, given numbers are marked in red, intermediate results are marked in yellow, and the final output is marked in green.

A useful exercise is to write this chain of operations as a single arithmetic expression:

$$((8 + 8) \times 4 + 2) \times 9 = 594.$$

This game is a valuable activity for refining arithmetic capabilities and training children in mental calculation within the range of three-digit numbers. In the lower grades, it is possible to simplify the game by using a two-digit target number, along with one-digit input numbers. One could also restrict the operations to, say, addition and subtraction, or addition and multiplication (or even just one operation).

Below are additional challenges that can be tried out in class:

(1) Use the numbers 4, 5, 2, 6, 9 and 7 to reach the target number 443. A possible solution is

$$9 \times 7 \times (5 + 2) + 6 - 4 = 443.$$

(2) Use the numbers 8, 10, 75, 50, 7 and 4 to reach the target number 668. A possible solution is

$$(4 + 8) \times 50 + 75 - 7 = 668.$$

(3) Use the numbers 8, 2, 75, 9, 10 and 4 to reach the target number 657. A possible solution is

$$(75 - 2) \times 9 = 657.$$

Mathematical problems

Problem 21.1. Draw the computation network corresponding to the following arithmetic expression, and evaluate it:

$$6 + 18 \div 2 \times 3 - 1 + 54 \div 4 \div 2.$$

Problem 21.2. Draw the computation network corresponding to the following arithmetic expression, and evaluate it:

$$6 + 18 \div 2 \times (3 - 1) + 54 \div (4 \div 2).$$

Problem 21.3. Play the *digits and letters* game. Use the numbers 1, 2, 3 and 5 to reach the target number 24. Two possible solutions are

$$(2 + 3) \times 5 - 1 = 24$$
$$(2 \times 3) \times (5 - 1) = 24.$$

Find two additional solutions.

Problem 21.4. Reach each of the target numbers

$$0, 1, 2, 3, 4, 6, 7, 8, 9, 15, 16, 17$$

using the numbers 4, 4, 4 and 4. You must use all the numbers. For example, you can reach the target number 2 as follows:

$$4 \div 4 + 4 \div 4 = 2.$$

Problem 21.5. Delete as many parentheses as possible in the following arithmetic expression, without changing the order of operations:

$$(((5 \times (3 + 4)) \div 7) + (63)) \times 2 - (5 \times 6).$$

Problem 21.6. Add one or more pairs of parentheses in order to obtain a valid equation:

$$6 + 24 \div 6 \times 3 = 15.$$

Chapter 22

Division with a Remainder

In Chapters 8 and 20, we studied the division operation and learned to evaluate quotients in cases where the two operands and the quotient are all integers. In general, the quotient of two integers may not be an integer. There are two possible ways to cope with division problems in which the quotient is not an integer. The first involves the notion of a **remainder**—this is the focus of the present chapter. The second involves the notion of **fractions**, and this approach will be used from Chapter 27 onward.

Hurdles:

(a) This is the first encounter with arithmetic expressions that are not equal to a whole number.
(b) The role of the remainder in word problems may differ from one problem to another.

Remedies:

(a) Emphasize the distinct meanings of the quotient and the remainder; interpret a non-zero remainder as indicating that division has not been completed.
(b) Expose children to a variety of word problems in which remainders assume different roles.

Opportunities:

(a) Division with a remainder is an opportunity to introduce fractions through the backdoor; a non-zero remainder occurs in cases where we are unable to "break a whole".

(b) The arithmetic of remainders is a fascinating extra-curricular topic.

22.1 Remainders

Consider the arithmetic expression:

$$25 \div 3.$$

It value, as expounded in Chapter 8, can be interpreted in two alternative ways:

(1) **Partitive division**: 25 apples where divided evenly into 3 bags. The arithmetic expression $25 \div 3$ represents the number of apples in each bag.
(2) **Quotative division**: 25 apples were divided into bags, 3 apples in each bag. The arithmetic expression $25 \div 3$ represents the number of bags needed.

Examine the partitive division interpretation: when partitioning 25 apples evenly into 3 bags, we end up placing 8 apples in each bag, with one apple leftover.

Examine then the quotative division interpretation: when partitioning 25 apples into triads (sets of 3 items), we end up forming 8 triads, with one apple leftover.

In both concrete interpretations of the abstract arithmetic expression $25 \div 3$, we have 1 apple leftover that will not go into any bag. If not for this leftover apple, that is, if there were only 24 apples at the start, the quotient would be 8—in the partitive division context each of the bags would contain 8 apples, and in the quotative division context 8 bags would be needed.

But the question prevails: what value should we attribute to the arithmetic expression $25 \div 3$? What should we do with the remaining apple?

There are two possible ways to handle the remaining apple. We will demonstrate both of them using the partitive division approach:

(1) Cut the remaining apple into 3 equal parts and place each piece, which we call a **third** of an apple, into another bag. In this case, each of the 3 bags contains 8 apples plus a third of an apple.
(2) Leave the remaining apple undivided. In this case, each of the 3 bags contains 8 apples, and there remains 1 apple that has not been divided. We call this apple a **remainder**.

The first alternative uses **fractions**, which quantify parts of a whole. This will be the approach we will adopt from Chapter 27 onward. In a number system restricted to whole numbers, only the second alternative prevails, in which case, the outcome of division is expressed by two integers—a quotient and a remainder:

The value of the arithmetic expression $25 \div 3$ is 8 and a remainder of 1.

The existence of a remainder signifies that division has not been completed. The remainder represents that part of the dividend that could not be divided without breaking a whole into parts.

22.1.1 Notation

Every mathematical entity or concept has its own dedicated notation, and so should have the result of division with a remainder. In the United States, the standard notation for an expression combining a quotient and a remainder is to separate the two numbers by the letter "R" (short for "remainder"). For example:

$$25 \div 3 = 8\,\mathrm{R}\,1.$$

This notational convention has, however, a shortcoming, which I will describe now.

Consider the two arithmetic expressions:

$$7 \div 3 \qquad \text{and} \qquad 9 \div 4.$$

Using the standard notation for division with a remainder, their values are

$$7 \div 3 = 2\,\mathrm{R}\,1 \qquad \text{and} \qquad 9 \div 4 = 2\,\mathrm{R}\,1,$$

namely, both seem to have the same value. By the **transitivity** of the equality relation, this would imply that

$$7 \div 3 = 9 \div 4.$$

This, however, is quite a dubious equality, which becomes strictly incorrect the moment we augment our number system with fractions.

The reason for this ambiguity can be attributed to a deficient system of notations. The remainder of 1 in the identity

$$7 \div 3 = 2\,\mathrm{R}\,1,$$

represents a remaining item that ought to be divided into 3 equal parts. On the other hand, the remainder of 1 in the identity

$$9 \div 4 = 2\,\mathrm{R}\,1,$$

represents a remaining item that ought to be divided into 4 equal parts. The standard notation for the remainder fails to capture an important aspect of "incomplete division"— into how many parts were we supposed to partition that remainder?

I am going to propose an alternative notation for quotients and remainders. To indicate explicitly that the remainder is a quantity that ought to be divided by the same divisor as the original dividend, we will use the following notation:

$$25 \div 3 = 8 + \mathrm{R}\,(1 \div 3).$$

In words: the outcome of dividing 25 into 3 is a quotient of 8 plus a remainder of 1 that should have been divided into 3.

The immediate advantage of this new notation is that it no longer suggests that $7 \div 3$ equals $9 \div 4$, as

$$7 \div 3 = 2 + \mathrm{R}\,(1 \div 3) \qquad \text{and} \qquad 9 \div 4 = 2 + \mathrm{R}\,(1 \div 4).$$

Another advantage of this notation is that it paves the way toward using fractions to express quotients of integers. At this stage, $\text{R}(1 \div 3)$ is only a notation for the remainder. Later on, once the number system has been extended to include fractions, $1 \div 3$ is a number—a third, and then the identify

$$25 \div 3 = 8 + \text{R}(1 \div 3)$$

acquires a new meaning, as an equivalence between two numerical expressions. Children who adopt the proposed notation for division with a remainder may find it easier later on to internalize division resulting with fractions.

22.1.2 The range of the remainder

Consider the arithmetic expression:

$$24 \div 6.$$

Its value is 4. Interpreted in terms of **partitive division**, if 24 items are divided equally into 6 shares, there are 4 items in each share. Interpreted in terms of **quotative division**, if 24 items are partitioned into quotas comprising 6 items each, there will be 4 such quotas. In both cases, there is no remainder, or equivalently, we may say that the remainder is zero. In our notation,

$$24 \div 6 = 4 + \text{R}(0 \div 6).$$

Such a notation may seem unnecessarily cumbersome; why would we ever want to denote a non-existing remainder? As will be seen below, remainders fulfill interesting arithmetic properties, which are better understood if we account explicitly for the possibility of the remainder being equal to zero.

Next, consider the arithmetic expression,

$$138 \div 7,$$

but don't try to evaluate it. What can be said about the remainder? For example, could the remainder be equal to 9?

To answer this question, let's interpret the remainder of this expression in terms of quotative division:

> Some 138 nuts were divided onto plates, each plate containing 7 nuts. The remainder of the expression $138 \div 7$ is the number of nuts remaining when there are fewer nuts remaining than needed to fill one more plate.

Even without calculating anything, we infer that this remainder must comprise fewer than 7 nuts, otherwise we could have filled at least one more plate with 7 nuts. That is, the remainder is always a whole number between zero (if there is no remainder) and 6 (the largest possible remainder).

Instead, if we had considered the arithmetic expression $2315 \div 16$, and interpreted the remainder as the number of nuts remaining undivided when 2315 nuts are placed onto plates, each containing 16 nuts, we would have inferred that this remainder is a whole number between zero (no remainder) and 15 (the largest possible remainder—one less that the divisor).

Thus, we may formulate the following conclusion:

> The remainder is a whole number greater or equal to zero and less than the divisor.

22.1.3 Inverse multiplication equation

In our first encounter with division (Volume 1, p. 99), we described division as an operation inverse to multiplication. The value of the arithmetic expression $a \div b$ is the solution to an unknown-factor equation

$$\boxed{?} \times b = a.$$

Since multiplication is commutative, it is also the solution to the unknown-factor equation

$$b \times \boxed{?} = a.$$

For example,

$$24 \div 4 = \boxed{6},$$

because

$$4 \times \boxed{6} = 24 \qquad \text{or because} \qquad \boxed{6} \times 4 = 24.$$

In this section we will see how division relates to multiplication in the presence of a remainder.

Consider once again our running example:

$$25 \div 3 = 8 + \mathrm{R}\,(1 \div 3),$$

along with the two types of word problems it could model:

(1) **Partitive division**: When dividing 25 items into 3 sets, we obtain 8 items in each set, plus 1 item left over. A number sentence describing this partition of the 25 items is

$$3 \times 8 + 1 = 25.$$

Please note: In division with a remainder there are *two* unknowns—a quotient and a remainder. The quotient represents the number of items in each set—8 in this case—and the remainder is the number of items that remain undivided—1 in this case. In order to differentiate between the two unknowns, we marked the quotient in orange and the remainder in blue.

(2) **Quotative division**: When dividing 25 items into sets comprising 3 items each, we obtain 8 sets, plus 1 item left over. A number sentence describing this partition of the 25 items is

$$8 \times 3 + 1 = 25.$$

Thus, when expressing the value of $25 \div 3$ in terms of a quotient and a remainder, there are two unknowns,

$$25 \div 3 = \boxed{?} + \mathrm{R}\,(\boxed{?} \div 3),$$

which are the solutions to the unknown-multiplicand equation,

$$3 \times \boxed{?} + \boxed{?} = 25,$$

or, alternatively, the unknown-multiplier equation,

$$\boxed{?} \times 3 + \boxed{?} = 25.$$

Please note: such an equation with two unknowns has multiple solutions, for example,

$$3 \times \boxed{1} + \boxed{22} = 25 \qquad \text{and} \qquad 3 \times \boxed{4} + \boxed{13} = 25,$$

which would suggest that

$$25 \div 3 = \boxed{1} + \mathrm{R}\,(\boxed{22} \div 3) \qquad \text{and} \qquad 25 \div 3 = \boxed{4} + \mathrm{R}\,(\boxed{13} \div 3).$$

In a sense, these identities are not totally erroneous. If you divide 25 items into three sets, you may well distribute 1 items to each set, remaining with 22 items, or distribute 4 items to each set, remaining with 13. In division with a remainder, it is implicitly assumed that the quotient is the *maximum* number of items that can be distributed to each set, hence the remainder does not exceed the divisor—it can only be 0, 1 or 2. Under these conditions, the *unique* solution to the unknown-factor equation is

$$3 \times \boxed{8} + \boxed{1} = 25,$$

which implies that

$$25 \div 3 = \boxed{8} + \mathrm{R}\,(\boxed{1} \div 3).$$

More generally, let a and b denote integers and consider the expression $a \div b$. In division with a remainder, the quotient and the remainder, $\boxed{?}$ and $\boxed{?}$, are the unique solutions to the equation with unknowns,

$$b \times \boxed{?} + \boxed{?} = a,$$

where $\boxed{?}$ is a whole number between 0 and $b - 1$.

The modulo operation

There exists a binary operation, **modulo**, which returns the remainder obtained when the first operand is divided by the second operand. This operation is commonly denoted by mod. Thus, for example,

$$24 \bmod 3 = 1,$$

because 1 is the remainder obtained when dividing 24 by 3.

22.2 Word problems

The division operation models situations where we partition items into sets of equal size. In this section we will inspect a variety of such situations, and see how the presence of a remainder manifests itself in each case.

Consider the word problem:

> 32 people are going on a trip. Each car has room for 5 people. How many cars are needed?

Apparently, this is a quotative division problem. The dividend is the number of people, 32, and the divisor is the number of riders in each car, 5. Thus, we need to find out how many groups of 5 people can be made from a group of 32 people. The corresponding arithmetic expression is

$$32 \div 5.$$

To evaluate this expression, we need to find a whole number quotient and a whole number remainder ranging between 0 and 4, such that

$$\boxed{?} \times 5 + \boxed{?} = 32.$$

The solution is not hard to find:

$$\boxed{6} \times 5 + \boxed{2} = 32,$$

that is,

$$32 \div 5 = \boxed{6} + \mathrm{R}(\boxed{2} \div 5).$$

But pay attention: we were asked a very specific question—"how many cars are needed?". The answer is 7, since 6 cars are not sufficient to accommodate 32 people, whereas 7 cars suffice (and in fact, could accommodate as many as 35 people). Yet, the answer is neither the quotient nor the remainder of the expression $32 \div 5$. The answer, 7, is the least number, which when multiplied by 5, yields a product that is at least 32.

Consider the next word problem:

> Marc Twain Elementary organizes a basketball tournament. Each team has 5 players. How many teams can the 32 fourth grade students send to the tournament?

Once again, we are facing a problem of quotative division. The dividend is the total number of children, 32, and the divisor in the number of children in each group, 5. We need to find out how many groups of 5 children can be formed with a group of 32 children. The corresponding arithmetic expression is once again

$$32 \div 5,$$

which equals, as we've already seen, $6 + \mathrm{R}(2 \div 5)$.

So how many teams can fourth grade send to the tournament? Only 6. A remainder of 2 children does not suffice to create another team. Unlike the previous question, here we had to find the *largest number*, which when multiplied by 5, yields a product not exceeding 32. Both the previous question and the present one exhibit division with a remainder, however, the presence of a non-zero remainder has a different implication in each situation.

Finally, consider one more word problem:

> A group of 5 hikers were asked to bring 32 liters of water. How many liters of water will each hiker carry if they divide the load evenly?

This time we are facing a problem of partitive division. The dividend is the total number of liters of water the hikers need to carry, 32. The divisor is the number of hikers, 5. We have to find out how many liters are in each share if we each divide the 32 liters into 5 equal shares. Once again, division with a remainder yields

$$32 \div 5 = 6 + \mathrm{R}\,(2 \div 5).$$

So how many liters will each hiker carry? 6? 7? If they fulfill the requirement of carrying a total of 32 liters and insist on equal loads, then they must divide among them the 2 remaining liters; each hiker has to carry 6 liters and two fifths of a liter. This type of solution requires familiarity with fractions, a concept that will be introduced in Chapter 27.

Please note: we examined three word problems, which, at first glance, are all related to the arithmetic expression $32 \div 5$. Nevertheless, the solution to the first problem was 7, the solution to the second problem was 6 and the solution to the third problem was 6 and 2/5. Only in one instance, the solution to the problem was the whole number quotient. In none of the problems was the value of the remainder requested.

One should not come to the hasty conclusion that the value of a remainder is never of interest. Consider the word problem:

> Gili had 32 pieces of candy which she wanted to divide evenly to 5 of her friends. How many pieces of candy did she keep to herself?

Once again, 32 items are divided into 5 groups, but this time, we are only asked about the remainder.

22.3 Evaluating division with a remainder

In Chapter 20, we learned to evaluate quotients of multi-digit numbers using a variety of methods, notably the long division algorithm. In this section, we apply long division in situations where the result includes a remainder.

Consider the expression:

$$80 \div 3.$$

We evaluate it stage-by-stage:

(a) **Divide tens**: We have to divide 8 tens by 3. Since 8 is not divisible by 3, we are looking for the largest number which, when multiplied by 3, yields a product not exceeding 8. This number is 2. We write 2 in the space allocated for the tens digit of the quotient, above the tens digit of the dividend.

(b) **Find out how many tens were divided**: We were supposed to divide 8 tens, however, we only divided as many tens as we could without decomposing tens into smaller decimal units. To find out how many tens were actually divided without having to memorize information from previous stages, we multiply the 2 tens in the quotient by the divisor, 3. The product is 6 tens. We write 6, representing the tens that were actually divided, under the 8 representing the tens that ought to be divided.

(c) **Calculate the remainder**: To find out how many tens have not yet been divided, we subtract the 6 tens that were actually divided from the 8 tens that ought to be divided. We obtain a remainder of 2 tens, which we write under both numbers.

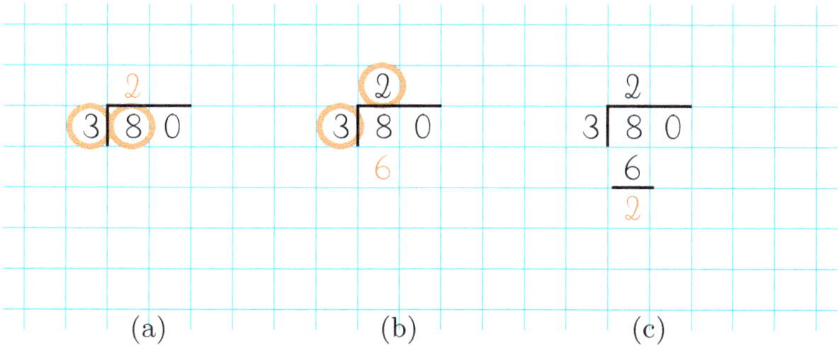

(a) (b) (c)

(d) **Decompose tens into ones**: We decompose the remaining 2 tens into 20 ones by "pulling down" the ones digit of the dividend, 0, to the right of the number of remaining tens. We obtain a two-digit number, 20, representing the number of ones that have yet to be divided.

(e) **Divide ones**: We have to divide 20 ones by 3. Since 20 is not divisible by 3, we are looking for the largest number, which when multiplied by 3, yields a product not exceeding 20. This number is 6. We write 6 in the space allocated for the ones digit of the quotient.

(f) **Find out how many ones were divided**: We were supposed to divide 20 ones, however, we only divided as many ones as we could without decomposing them into smaller units. To find out how many ones were actually divided, we multiply the 6 ones in the quotient by the divisor, 3. The product is 18 ones. We write 18, representing the ones that were actually divided, under the 20 representing the ones that ought to be divided.

(g) **Calculate the remainder**: To find out how many ones have not yet been divided, we subtract the 18 ones that were divided from the 20 ones that ought to be divided. We obtain a remainder of 2 ones, which we write, using our notation for the remainder, to the right of the quotient.

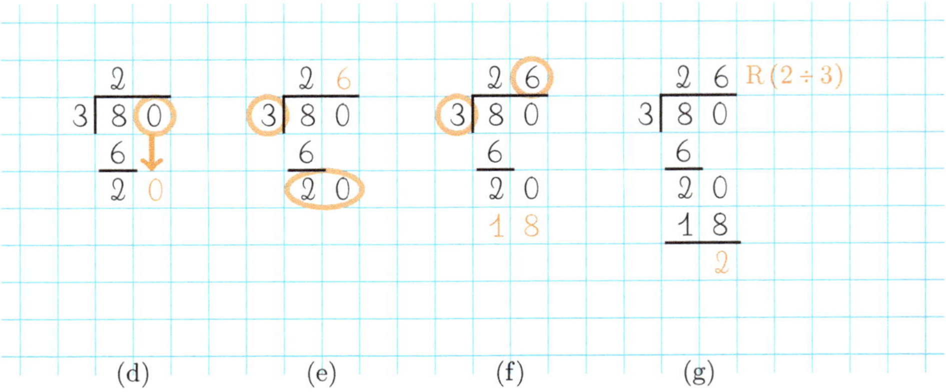

22.4 Arithmetic of remainders

22.4.1 Odd and even numbers

The concept of **parity**, which distinguishes between odd and even numbers, was introduced in Chapter 6. Recall the two different, yet equivalent definitions for the parity of a number:

Definition 22.1. A whole number is **even** if it can be represented as a sum of two equal whole numbers. A whole number that cannot be represented as a sum of two equal whole numbers is called **odd**.

Definition 22.2. A whole number is **even** if it can be represented as a multiple of 2, i.e., a repeated addition of 2. A whole number that cannot be represented as a multiple of 2 is called **odd**.

Now that we are acquainted with the concept of division with a remainder, we can provide a third equivalent definition to the parity of a number:

Definition 22.3. A whole number is **even** if it is divisible by 2 with a remainder of 0. It is **odd** if it is divisible by 2 with a remainder of 1.

Please note: the remainder of division by 2 is either 0 or 1, hence, every integer is either even or odd according to Definition 22.3.

In Volume 1, p. 62, we learned that parity satisfies arithmetic relations, which we called the **arithmetic of parity**. Specifically, it suffices to know the parity of two numbers to deduce the parity of their sum and their product. This fact is summarized by the following relations:

$$\text{even} + \text{even} = \text{even}$$
$$\text{odd} + \text{even} = \text{odd}$$
$$\text{odd} + \text{odd} = \text{even}$$
$$\text{even} \times \text{even} = \text{even}$$
$$\text{odd} \times \text{even} = \text{even}$$
$$\text{odd} \times \text{odd} = \text{odd}.$$

In view of Definition 22.3, we may reformulate those 6 relations as follows:

$$\begin{aligned}
\mathrm{R}(0 \div 2) + \mathrm{R}(0 \div 2) &= \mathrm{R}(0 \div 2)\\
\mathrm{R}(0 \div 2) + \mathrm{R}(1 \div 2) &= \mathrm{R}(1 \div 2)\\
\mathrm{R}(1 \div 2) + \mathrm{R}(1 \div 2) &= \mathrm{R}(0 \div 2)\\
\mathrm{R}(0 \div 2) \times \mathrm{R}(0 \div 2) &= \mathrm{R}(0 \div 2)\\
\mathrm{R}(0 \div 2) \times \mathrm{R}(1 \div 2) &= \mathrm{R}(0 \div 2)\\
\mathrm{R}(1 \div 2) \times \mathrm{R}(1 \div 2) &= \mathrm{R}(1 \div 2).
\end{aligned}$$

The first relation, for example, asserts that the sum of two numbers divisible by 2 with a remainder of 0 is divisible by 2 with a remainder of 0.

Another common, and more succinct notation for the same relations uses the modulo operation (see p. 68):

$$\begin{aligned}
0 + 0 &= 0 \pmod 2 \qquad & 0 \times 0 &= 0 \pmod 2\\
0 + 1 &= 1 \pmod 2 \qquad & 0 \times 1 &= 0 \pmod 2\\
1 + 1 &= 0 \pmod 2 \qquad & 1 \times 1 &= 1 \pmod 2,
\end{aligned}$$

where the suffix (mod 2) indicates that the equality has to be interpreted in the sense of remainders of division by 2; see also the addition and multiplication tables pertinent to the arithmetic of parity in Volume 1, pp. 73,74.

The fact that we can generate addition and multiplication tables for the arithmetic of parity (or modulo-2 arithmetic), reflects the fact that the parity of a sum or a product is unambiguously determined by the parity of the two operands. A natural question is whether an arithmetic of remainders can be developed for divisors other than 2. The answer is positive: an arithmetic of remainders can be formulated for any integer divisor. In the next section we develop, as an example, modulo-3 arithmetic, i.e., the arithmetic of remainders of division by 3.

22.4.2 Modulo-3 classification of integers

As expounded on p. 66, there are three possible outcomes when dividing an integer by 3:

(a) The remainder is 0.
(b) The remainder is 1.
(c) The remainder is 2.

Consider first integers divisible by 3 with no remainder. Just as there were two equivalent definitions for the evenness of a number—one using a partitive division viewpoint and another using a quotative division viewpoint—so there are two equivalent definitions for a number being divisible by 3 with no remainder:

Definition 22.4. An whole number a is divisible by 3 with no remainder, i.e., $a \bmod 3 = 0$, if it can be represented as a sum of three equal whole numbers.

Definition 22.5. A whole number a is divisible by 3 with no remainder, i.e., $a \bmod 3 = 0$, if it can be represented as a multiple of 3, or as a repeated addition of 3's.

For example, according to Definition 22.4, 18 is divisible by 3 with no remainder because it is possible to express it as

$$18 = 6 + 6 + 6.$$

According to Definition 22.5, it is divisible by 3 with no remainder because it is possible to express it as

$$18 = 3 + 3 + 3 + 3 + 3 + 3.$$

Moreover, Definition 22.5 suggests a systematic way of enumerating all of those integers divisible by 3 with no remainder: start at 0 and skip-count by 3,

$$0, 3, 6, 9, 12, 15, 18, \cdots$$

Likewise, is it possible to characterize in two equivalent ways all of those numbers which are divisible by 3 with a remainder of 1:

Definition 22.6. A whole number a is divisible by 3 with a remainder of 1, i.e., $a \bmod 3 = 1$, if it is the successor of a number that can be represented as a sum of three equal whole numbers.

Definition 22.7. A whole number a is divisible by 3 with a remainder of 1, i.e., $a \bmod 3 = 1$, if it is the successor of a number that can be represented as a multiple of 3.

For example, according to Definition 22.6, 25 is divisible by 3 with a remainder of 1 because it is possible to express it as

$$25 = 8 + 8 + 8 + 1.$$

According to Definition 22.7, it is divisible by 3 with a remainder of 1 because it is possible to express it as

$$25 = 3 + 3 + 3 + 3 + 3 + 3 + 3 + 3 + 1.$$

Moreover, Definition 22.7 suggests a systematic way of enumerating all of those integers divisible by 3 with a remainder of 1: start at 1 and skip-count by 3,

$$1, 4, 7, 10, 13, 16, 19, \cdots$$

Finally, is it possible to characterize all of those numbers which are divisible by 3 with a remainder of 2 in two equivalent ways:

Definition 22.8. A whole number a is divisible by 3 with a remainder of 2, i.e., $a \bmod 3 = 2$, if it is two more than a number that can be represented as a sum of three equal whole numbers.

Definition 22.9. A whole number a is divisible by 3 with a remainder of 2, i.e., $a \bmod 3 = 2$, if it is two more than a number that can be represented as a multiple of 3.

For example, according to Definition 22.8, 14 is divisible by 3 with a remainder of 2 because it is possible to express it as

$$14 = 4 + 4 + 4 + 2.$$

According to Definition 22.9, it is divisible by 3 with a remainder of 2 because it is possible to express it as

$$14 = 3 + 3 + 3 + 3 + 2.$$

To systematically enumerate all of those integers which are divisible by 3 with a remainder of 2, we start at 2 and skip-count by 3:

$$2, 5, 8, 11, 14, 17, 20, \cdots.$$

We have already seen that odd and even numbers occur in alternate order on the number line. A similar phenomenon occurs with remainders of division by 3. On the number line below, the numbers have been colored according to their remainder when divided by 3. Numbers divisible by 3 with no remainder are colored red, numbers divisible by 3 with a remainder of 1 are colored blue, and numbers divisible by 3 with a remainder of 2 are colored yellow. Like odd and even numbers, the modulo-3 value of numbers alternates on the number line.

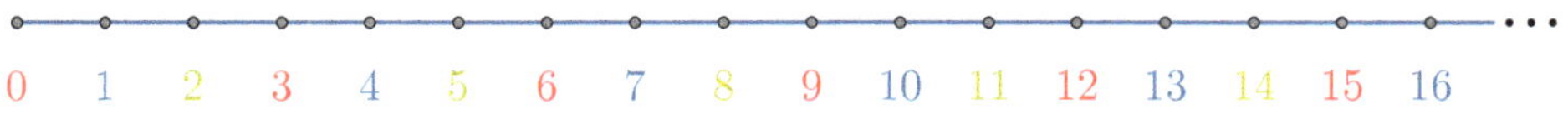

22.4.3 Modulo-3 addition

The sum of two even numbers is always even. Do similar properties hold for remainders of division by 3? More formally, let a and b be two integers, and suppose we know the values of

$$a \bmod 3 \qquad \text{and} \qquad b \bmod 3,$$

but we do not know what a and b are. Is this information sufficient to infer the value of

$$(a + b) \bmod 3?$$

The answer to this question turns out to be positive.

Let's start examining the sum of two numbers divisible by 3 with no remainder. If a and b are both divisible by 3 with no remainder, then according to Definition 22.5, they both can be expressed as repeated additions of 3,

$$a = 3 + 3 + \cdots + 3 + 3 \qquad \text{and} \qquad b = 3 + 3 + \cdots + 3 + 3,$$

where the ellipses indicate that we do not know a priori how many times this repeated addition occurs.

An illustration of a set comprising a number of items divisible by 3 with no remainder is shown below: the items can be grouped into triples,

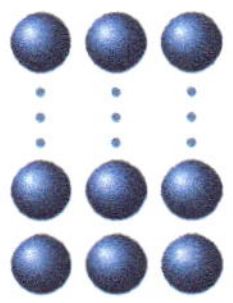

If we join two such sets, we obtain a compound set that is again a collection of triples,

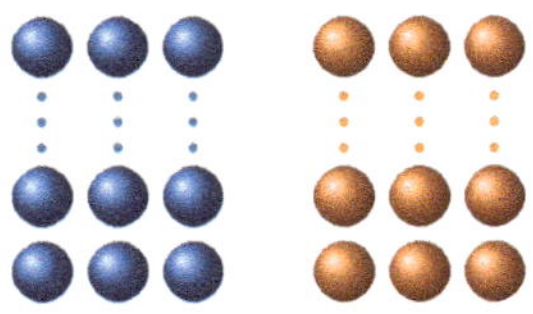

In mathematical notation:

$$(3 + 3 + \cdots + 3) + (3 + 3 + 3 + \cdots + 3) = 3 + 3 + 3 + \cdots + 3.$$

In other words, we have just shown that

$$a \bmod 3 = 0 \quad \text{and} \quad b \bmod 3 = 0 \quad \text{implies} \quad (a + b) \bmod 3 = 0,$$

or

$$0 + 0 = 0 \qquad (\bmod 3).$$

Consider another example: suppose that a and b are both divisible by 3 with a remainder of 2, i.e.,

$$a \bmod 3 = 2 \qquad \text{and} \qquad b \bmod 3 = 2.$$

According to Definition 22.9, they both can be expressed as two plus a repeated addition of 3,

$$a = 3 + 3 + \cdots + 3 + 3 + 2 \qquad \text{and} \qquad b = 3 + 3 + \cdots + 3 + 3 + 2.$$

An illustration of a set comprising a number of items divisible by 3 with a remainder of 2 is shown below:

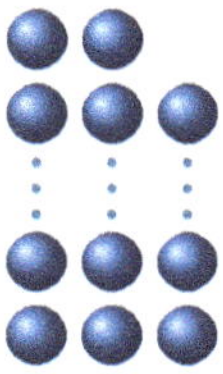

If we join two such sets, we obtain a set that can be organized as a collection of triples plus *one* remaining item:

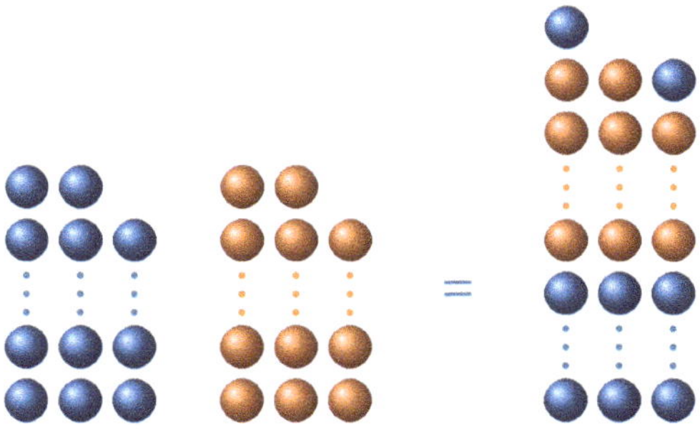

In mathematical notation:

$$\begin{aligned}(3 + 3 + \cdots + 3 + 2) + (3 + 3 + 3 + \cdots + 3 + 2) &= 3 + 3 + 3 + \cdots + 3 + 4 \\ &= 3 + 3 + 3 + \cdots + 3 + 3 + 1.\end{aligned}$$

In other words, we have just shown that

$$a \bmod 3 = 2 \quad \text{and} \quad b \bmod 3 = 2 \quad \text{implies} \quad (a + b) \bmod 3 = 1,$$

or

$$2 + 2 = 1 \qquad (\bmod 3).$$

Similar considerations for all possible remainders of division by 3 lead to the following list of addition facts:

$$\begin{array}{llll} 0+0=0 & (\text{mod}\,3) & 1+1=2 & (\text{mod}\,3) \\ 0+1=1 & (\text{mod}\,3) & 1+2=0 & (\text{mod}\,3) \\ 0+2=2 & (\text{mod}\,3) & 2+2=1 & (\text{mod}\,3), \end{array}$$

where the suffix (mod 3) indicates that the equality has to be interpreted in the sense of remainders of division by 3. These addition facts can also be arranged into a **modulo-3 addition table**,

+	0	1	2
0	0	1	2
1	1	2	0
2	2	0	1

More on modulo-3 addition

(1) Modulo-3 addition satisfies the **commutative property**, as apparent from the symmetry of the addition table (cf. Volume 1, p. 45).

(2) Modulo-3 addition satisfies the **associative property**. The associative property is harder to discern than the commutative property in the addition table. In principle one should verify that the associative property holds in all instances; for example,

$$\underbrace{\underbrace{(2+2)}_{1}+1}_{2} = \underbrace{2+\underbrace{(2+1)}_{0}}_{2}.$$

(3) Zero is the **identity element** of modulo-3 addition. That is, the sum of a number divisible by 3 with no remainder and another number is divisible by 3 with the same remainder as the second addend.

22.4.4 Modulo-3 multiplication

The product of two even numbers is always even. Do similar properties hold for remainders of division by 3? More formally, let a and b be two integers, and suppose we know the values of

$$a \bmod 3 \qquad \text{and} \qquad b \bmod 3,$$

but we do not know what a and b are. Is this information sufficient to infer the value of

$$(a \times b) \bmod 3?$$

The answer to this question turns out to be positive as well. However, it is less apparent than in the case of addition.

When facing a non-trivial question, it is always good practice to first examine particular examples. Take for example products of numbers divisible by 3 with remainders of 2. Two such numbers are

$$5 = 3 + 2 \qquad \text{and} \qquad 8 = 3 + 3 + 2.$$

Their product, 40, is divisible by 3 with a remainder of 1. The question is whether the same remainder would have been obtained had we taken factors other than 5 and 8.

Two other numbers divisible by 3 with remainders of 2 are,

$$11 = 3 + 3 + 3 + 2 \qquad \text{and} \qquad 8 = 3 + 3 + 2.$$

Their product, 88, is also divisible by 3 with a remainder of 1. Examining a few more examples, we can see that this seems to always be the case, namely,

$$a \bmod 3 = 2 \quad \text{and} \quad b \bmod 3 = 2 \quad \text{implies} \quad (a \times b) \bmod 3 = 1.$$

We will now elucidate the principles behind this rule.

Recall that products are conveniently visualized using **rectangular arrays**. The product of two numbers divisible by 3 with a remainder of 2 can be represented by the following rectangular array:

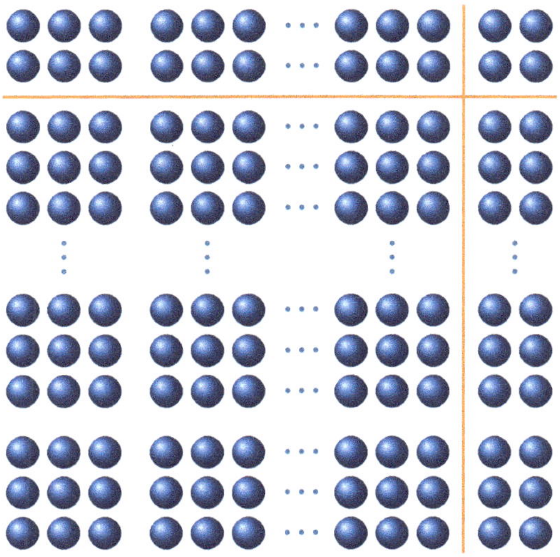

The spacing between the dots forming the array has been set to highlight the fact that both the number of rows and the number of columns are a sum of triples plus two.

The two orange lines partition the array into four sub-arrays:

(1) Lower-left: the number of rows is divisible by 3 with no remainder, hence the total number of dots is divisible by 3 with no remainder.
(2) Upper-left: the number of rows is divisible by 3 with no remainder, hence the total number of dots is divisible by 3 with no remainder.
(3) Lower-right: the number of columns is divisible by 3 with no remainder, hence the total number of dots is divisible by 3 with no remainder.
(4) Upper-right: contains 4 dots.

It follows that the total number of dots in the array is of the form

$$\underbrace{(3+\cdots+3)}_{\text{lower-left}}+\underbrace{(3+\cdots+3)}_{\text{upper-left}}+\underbrace{(3+\cdots+3)}_{\text{lower-right}}+\underbrace{4}_{\text{upper-right}}=3+\cdots+3+1,$$

thus revealing why it is always the case that

$$a \bmod 3 = 2 \quad \text{and} \quad b \bmod 3 = 2 \quad \text{implies} \quad (a \times b) \bmod 3 = 1.$$

Repeating similar considerations for all possible remainders of division by 3, we obtain the following list of multiplication facts:

$$\begin{array}{llll} 0 \times 0 = 0 & (\bmod 3) & 1 \times 1 = 1 & (\bmod 3) \\ 0 \times 1 = 0 & (\bmod 3) & 1 \times 2 = 2 & (\bmod 3) \\ 0 \times 2 = 0 & (\bmod 3) & 2 \times 2 = 1 & (\bmod 3), \end{array}$$

These multiplication facts can also be arranged into a **modulo-3 multiplication table**:

×	0	1	2
0	0	0	0
1	0	1	2
2	0	2	1

More on modulo-3 multiplication

(1) Modulo-3 multiplication satisfies the **commutative property**, as apparent in the symmetry of the multiplication table.
(2) Modulo-3 multiplication satisfies the **associative property**.

(3) One is the **identity element** of modulo-3 multiplication. That is, the product of a number divisible by 3 with a remainder of 1 and another number is divisible by 3 with the same remainder as the second factor.
(4) Zero is the **null element** of modulo-3 multiplication. That is, the product of a number divisible by 3 with no remainder and any another number is divisible by 3 with no remainder.
(5) Modulo-3 addition and modulo-3 multiplication satisfy the **distributive property**.

Mathematical problems

Problem 22.1. Determine the following quotients and remainders:

$$30 \div 4 = \boxed{?} + \mathrm{R}(\boxed{?} \div 4)$$
$$31 \div 4 = \boxed{?} + \mathrm{R}(\boxed{?} \div 4)$$
$$32 \div 4 = \boxed{?} + \mathrm{R}(\boxed{?} \div 4)$$
$$33 \div 4 = \boxed{?} + \mathrm{R}(\boxed{?} \div 4)$$
$$34 \div 4 = \boxed{?} + \mathrm{R}(\boxed{?} \div 4)$$
$$35 \div 4 = \boxed{?} + \mathrm{R}(\boxed{?} \div 4)$$
$$36 \div 4 = \boxed{?} + \mathrm{R}(\boxed{?} \div 4)$$
$$37 \div 4 = \boxed{?} + \mathrm{R}(\boxed{?} \div 4).$$

What regularity is revealed in this sequence of identities?

Problem 22.2. Consider the following three word problems:

- Alon solved 4 arithmetic problems in one hour. How long will it take him to solve 25 problems if he keeps working at the same pace?
- One taxicab seats 4 passengers. How many taxicabs are needed to transport 25 passengers?
- One notebook costs $4. How many notebooks can be purchased with $25?

(a) Solve all three problems.
(b) What characteristic is common to all three problems?
(c) What is the difference between them?
(d) How are those questions connected (if at all) to the concept of a remainder?

Problem 22.3. Write a division problem whose solution is "4 with a remainder of 7". Can you write more than one such problem? Can you write more than 7 such problems?

Problem 22.4. Solve the following equations with unknowns; if an equation has more than one solution, find all possible solutions.

$$\boxed{?} \div 4 = 5 + \mathrm{R}(\boxed{?} \div 4)$$
$$18 \div \boxed{?} = 3 + \mathrm{R}(\boxed{?} \div \boxed{?})$$
$$51 \div \boxed{?} = 3 + \mathrm{R}(\boxed{?} \div \boxed{?})$$
$$35 \div \boxed{?} = \boxed{?} + \mathrm{R}(2 \div \boxed{?}).$$

Problem 22.5. Consider a building in which apartment 1 is located on the first floor, and there are 3 apartments on each floor. On what floor is apartment 28? How would the answer change if there were 4 apartments on each floor? How is this question related to division? What is the significance of the remainder?

Problem 22.6. In this chapter, we examined the arithmetic of remainders of division by 3, and constructed addition and multiplication tables for modulo-3 arithmetic. Is it possible to define modulo-3 subtraction? Evaluate, for example, the following modulo-3 differences:

$$2 - 1 = \boxed{?} \pmod 3$$
$$0 - 1 = \boxed{?} \pmod 3.$$

Problem 22.7. Consider modulo-7 arithmetic:

(a) What are all of the numbers divisible by 7 with a remainder of 4?
(b) Construct a modulo-7 addition table.
(c) Construct a modulo-7 multiplication table.

Problem 22.8. Consider the modulo-3 addition table on p. 78. What common characteristics does it have with the "standard" addition table in Volume 1, p. 46?

Problem 22.9. Consider the modulo-3 multiplication table on p. 80. What common characteristics does it have with the "standard" multiplication table in Volume 1, p. 92?

Problem 22.10. Determine which of the following statements are correct:

(a) Any number divisible by 6 with a remainder of 2 is also divisible by 3 with a remainder of 2.
(b) Any number divisible by 3 with a remainder of 2 is also divisible by 6 with a remainder of 2.
(c) There exists a number divisible by 3 with a remainder of 2 and also divisible by 6 with a remainder of 2.
(d) There exists a number divisible by 5 with a remainder of 2 and also divisible by 6 with a remainder of 2.

Problem 22.11. If the current time is 6:00AM. What will be the time in 120 hours? What will be the time in 84 hours? What will be the time in 2405 hours? (Hint: modulo-24 arithmetic.)

Problem 22.12. December 31, 2011 falls on a Thursday. Based on this fact, find what day does December 31 fall on in 2010 and in 2012. Do not forget to take leap years into account! (Hint: modulo-7 arithmetic.)

Chapter 23

Prime and Composite Numbers

The natural numbers, $1, 2, 3, \ldots$, are classified into two subsets: **prime numbers** and **composite numbers**. A natural number is composite if it can be decomposed into a product of two smaller natural numbers; otherwise it is called prime. Learning the concepts of prime and composite numbers is essential for gaining a deep understanding of the number system and the arithmetic operations.

Opportunities:

(a) Testing for primality is an excellent opportunity to practice multiplication and division in a meaningful context.
(b) Prime number factorization is an important skill in the context of fractions.
(c) The study of prime numbers is an opportunity to expose children to real mathematical gems, including two of the world's most famous open problems.

23.1 Multiples and divisors

The following two definitions will be at the heart of the present chapter, and of the two subsequent chapters:

Definition 23.1. A natural number n is called a **multiple** of a natural number a, if n is the product of a and some other natural number. In other words, n is a multiple of a if there exists a natural number k, such that

$$n = k \times a.$$

For example, the natural number 24 is a multiple of the natural number 6 because 24 is the product of 6 and a natural number, 4,

$$24 = 4 \times 6.$$

Definition 23.2. A natural number a is called a **divisor** of natural number n if n is divisible by a with no remainder. In other words, a is a divisor of n if there exists a natural number k, such that

$$n = k \times a.$$

For example: the natural number 6 is a divisor of the natural number 24 because 24 is divisible by 6 with no remainder, or equivalently, because there exists a natural number, 4, such that

$$24 = 4 \times 6.$$

On multiples and divisors

(a) The term **factor** is a synonym for divisor.

(b) The concepts of multiple and divisor are reciprocal. If n is a multiple of a, then a is necessarily a divisor of n, and vice versa, if a is a divisor of n, then n is necessarily a multiple of a. For example, if it is known that 24 is a multiple of 6, then we can infer that 6 is a divisor of 24, and vice versa, if it is known that 13 is a divisor of 78, then we can infer that 78 is a multiple of 13.

(c) We introduce a relation sign $\lhd$ to denote the relationship between multiples and divisors; we call this sign the **divisor-of sign**. For example, the fact that 6 is a divisor of 24 is denoted as follows:

$$6 \lhd 24.$$

(d) Every natural number n is its own multiple because $n = 1 \times n$. From the reciprocal relation between multiples and divisors it follows that every natural number is also its own divisor. That is, for every natural number n,

$$n \lhd n.$$

(e) Every natural number n has an infinite number of multiples, the least of them being n itself. For example, 6 has an infinite number of multiples, which can be listed by the repeated addition of 6,

$$6, 12, 18, 24, 30, \ldots.$$

(f) The natural number 1 is a divisor of every natural number n because $n = n \times 1$, or equivalently, because every natural number is divisible by 1 with no remainder. That is, for every natural number n,

$$1 \lhd n.$$

(g) It follows that every natural number n greater than 1 has at least two divisors. The number 1 is the smallest divisor of every natural number, and n itself is the largest divisor of n. Natural numbers may have more than two divisors. For example, the divisors of 6 are 1, 2, 3 and 6.

23.1.1 Properties of multiples and divisors

Multiples and divisors have properties deriving directly from their definition.

23.1.1.1 Transitivity of the multiple-of relation

The multiple-of relation satisfies the following property:

If b is a multiple of a and c is a multiple of b, then c is a multiple of a.

This property can be expressed using the divisor-of relation sign.

$$\text{If} \quad a \lhd b \quad \text{and} \quad b \lhd c \quad \text{then} \quad a \lhd c.$$

For example, if we know that 38 is a multiple of 19, and we know that 114 is a multiple of 38, then we can infer that 114 is a multiple of 19. This property of the multiple-of relation is called **transitivity** (for more examples of transitivity see Volume 1, pp. 6, 11 and 168).

The transitivity of the multiple-of relation can be explained as follows:

(1) If b is a multiple of a, then, by definition, there exists a natural number k such that

$$b = k \times a.$$

(2) Furthermore, if c is a multiple of b, then, by definition, there exists a natural number m such that

$$c = m \times b.$$

(3) Integrating these two facts and using the associative property of multiplication, it follows that:

$$c = m \times b = m \times (k \times a) = (m \times k) \times a.$$

Thus, c equals $m \times k$ times a, which implies that c is a multiple of a.

23.1.1.2 Transitivity of the divisor-of relation

Due to the reciprocity between multiples and divisors, the transitivity property of the multiple-of relation implies that the divisor-of relation is also transitive:

If a is a divisor of b and b is a divisor of c, then a is a divisor of c.

For example, if we know that 7 is a divisor of 35 and we know that 35 is a divisor of 315, then we can infer that 7 is a divisor of 315.

23.1.1.3 Complementary divisor

In the context of the division operation, we learned about the concept of an **adjoint division equation** (Volume 1, p. 108). For example, if

$$24 \div 3 = 8,$$

then it follows that

$$24 \div 8 = 3.$$

The reason why one equality implies the other is that both derive from the same multiplication relation, $8 \times 3 = 24$.

The important implication of this observation is that if the natural number a is a divisor of the natural number n, namely, if $n \div a$ yields a quotient with no remainder, then this quotient, $n \div a$, is also the divisor of n. We call the divisors a and $n \div a$ **complementary divisors** of n.

For example, knowing that 3 is a divisor of 24, we deduce that $24 \div 3 = 8$ is also a divisor of 24. The numbers 3 and 8 are complementary divisors of 24.

Please note: It is quite possible that the divisor of a number and its complementary divisor turn out to be the same number. For example, 5 is a divisor of 25 (because 25 is divisible by 5 with no remainder), and its complementary divisor is $25 \div 5 = 5$.

23.1.2 Enumerating the divisors of a given number

Consider, for example, the number 24. Since its largest divisor is 24, it cannot have an infinite number of divisors. Our goal in this section is to find *all* the divisors of a given number.

The most naive approach would be to consider all the natural numbers between 1 and 24, and verify one-by-one whether they divide 24 with no remainder. Such a systematic verification would reveal that the divisors of 24 are

$$1, 2, 3, 4, 6, 8, 12, 24.$$

We arranged the eight divisors of 24 in increasing order. Note that the smallest divisor, 1, is complementary to the largest divisor, 24; likewise, the second smallest divisor, 2, is complementary to the second largest divisor, 12, and so on, as illustrated below:

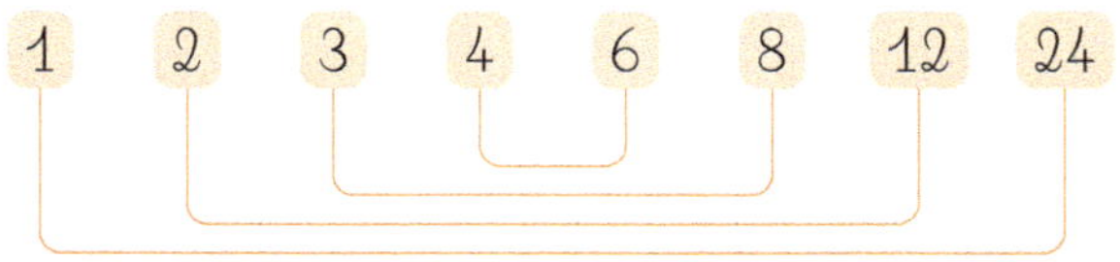

This observation suggests a more efficient method for finding all the divisors for 24:

(a) The number 1 is a divisor of 24, since it is a divisor of every natural number.

(b) The divisor complementary to 1 is $24 \div 1 = 24$.

(c) Is 2 a divisor of 24? Yes.

(d) The divisor complementary to 2 is $24 \div 2 = 12$. We deduce that 24 does not have divisors between 12 and 24, because if there were such a divisor, its complementary divisor would have been a whole number between 1 and 2. Of course, there is no such number.

(e) Is 3 a divisor of 24? Yes.

(f) The divisor complementary to 3 is $24 \div 3 = 8$. We deduce that 24 does not have a divisor between 8 and 12, because if there were such a divisor, its complementary divisor would have been a whole number between 2 and 3.

(g) Is 4 a divisor of 24? Yes.

(h) The divisor complementary to 4 is $24 \div 4 = 6$. We deduce that 24 does not have a divisor between 6 and 8.

(i) Is 5 a divisor of 24? No.

(j) The next natural number is 6. We already know that 6 is a divisor of 24, and we know that there are no divisors greater than 6 that have not yet been listed.

(k) To summarize, the divisors of 24 are 1, 2, 3, 4, 6, 8, 12 and 24.

If, for example, we were required to list all the divisors of 2048, we would repeat the same process, systematically finding every divisor and its complementary divisor. However, be aware that the larger the target number is, the lengthier is the process of enumerating all its divisors. In the following sections we will encounter even more efficient methods for finding systematically all the divisors of a given number, whether large or small.

23.2 Prime numbers

23.2.1 Prime and composite numbers

Examine all the divisors of the numbers between 1 and 20. Please pay attention to the following three facts, which we have already pointed out above:

(a) The number 1 is the only one in the list having a single divisor.

(b) All numbers n greater than 1 have at least two divisors: 1 and n.

(c) Some numbers have more than two divisors.

Number	Divisors
1	1
2	1,2
3	1,3
4	1,2,4
5	1,5
6	1,2,3,6
7	1,7
8	1,2,4,8
9	1,3,9
10	1,2,5,10

Number	Divisors
11	1,11
12	1,2,3,4,6,12
13	1,13
14	1,2,7,14
15	1,3,5,15
16	1,2,4,8,16
17	1,17
18	1,2,3,6,9,18
19	1,19
20	1,2,4,5,10,20

These observations lead us to the following definition:

Definition 23.3. A natural number having exactly two divisors is called a **prime number**. A natural number having more than two divisors is called a **composite number**.

As can be seen from the above table, the prime numbers up to 20 are:

$$2, 3, 5, 7, 11, 13, 17, 19.$$

The number 1 has a special status, for it is the only natural number that is neither prime nor composite (the reason for this exception will be explained below; see p. 94).

It is not known exactly when people first classified numbers into primes and composites. It is known, however, that prime numbers attracted considerable interest in ancient Egypt and Greece. For the past 2500 years, prime numbers have been at the focus of mathematical research. Even to this day, primes number pose great mysteries yet to be resolved.

23.2.2 The fundamental theorem of arithmetic

Prime numbers are viewed as the building blocks of all natural numbers, just as atoms are the building blocks of the material world. Prime numbers compose all the natural numbers in the following sense: every natural number is either a prime, or it can be represented as a product of prime numbers. For example, 20, which is a composite number, can be represented as the product of prime numbers:

$$20 = 2 \times 2 \times 5.$$

Thus, we can view the prime numbers as elementary building blocks with the multiplication operation being the "glue" connecting prime numbers to form composite numbers.

The figure below envisions the world of natural numbers as a factory whose raw material is prime numbers and whose product is all the natural numbers.

One of the most important theorems in the theory of numbers, known as **the fundamental theorem of arithmetic**, states that:

> Every natural number greater than 1 is either prime, or has a unique representation as a product of prime numbers.

Please note that the fundamental theorem of arithmetic makes two distinct statements:

(1) Every composite number can be represented as a product of prime numbers.

(2) There is only one unique collection of prime numbers whose product equals that number.

The first part of the theorem states, for example, that the numbers 96, 81 and 133, which are composite numbers, can be represented as products of prime numbers. Indeed,

$$96 = 2 \times 2 \times 2 \times 2 \times 2 \times 3$$
$$81 = 3 \times 3 \times 3 \times 3$$
$$133 = 7 \times 19.$$

The second part of the theorem states that for each of those numbers, there is no other representation as a product of prime numbers (other than interchanging the order of the factors).

Proving both parts of the fundamental theorem of arithmetic requires advanced skills. Yet, the first part can be proven by a relatively simple argument:

> **Claim 23.1.** *Every composite number n greater than 1 can be represented as a product of prime numbers.*

The argument is the following:

(a) Since n is composite, it can be represented as a product $n = a \times b$, where a and b are natural numbers different than 1 or n. In particular, both a and b must be smaller than n.

(b) If both a and b happen to be prime numbers, then the proof is complete, as we represented n as a product of two prime numbers.

(c) Otherwise, if either a or b are composite numbers, we can represent them as products of two natural numbers. For example, if a is a composite number, we can represent is as a product, $a = c \times d$, where both c and d are smaller than a. Thus, $n = (c \times d) \times b$.

(d) We proceed this way, decomposing each composite factor into a product of two smaller factors. This process is guaranteed to terminate as the factors cannot be reduced indefinitely. At the end of this process, we obtain a representation of n as a product of prime numbers.

Let's demonstrate the idea behind the proof by decomposing the composite number 72 into a product of prime numbers:

(a) Since 72 is a composite number, we can represent it as a product of two natural numbers other than 1 and 72. For example $72 = 8 \times 9$ (there are other possible factorizations of 72, such as 4×18).

(b) The factor 8 is a composite number, which we can represent as a product, $8 = 2 \times 4$, obtaining

$$72 = 2 \times 4 \times 9.$$

(c) The factor 4 is a composite number, which we can represent as a product, $4 = 2 \times 2$, obtaining

$$72 = 2 \times 2 \times 2 \times 9.$$

(d) The factor 9 is a composite number, which we can represent as a product, $9 = 3 \times 3$, obtaining

$$72 = 2 \times 2 \times 2 \times 3 \times 3.$$

(e) At this stage, all the factors on the right hand side are prime numbers, and we have thereby completed the factorization of 72 into a product of prime numbers.

Below is an illustration of this particular factorization of 72 into a product of prime numbers, using a **tree diagram** (in which the tree is in fact upside down—root on top and leaves at the bottom):

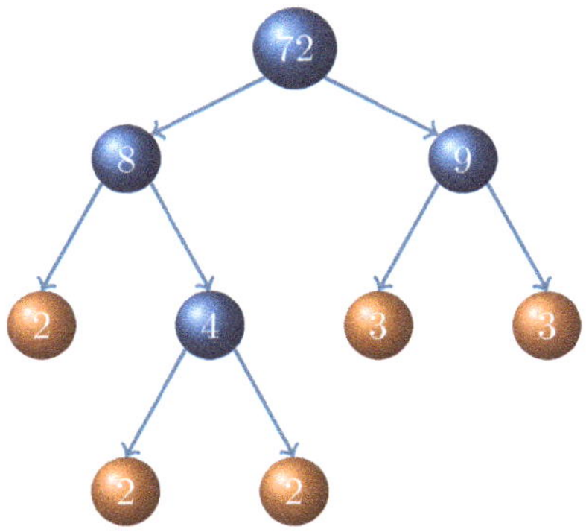

The **root** of the tree, 72, is the number that we want to decompose. Two **branches** descend from the root, each pointing to a **node**. The two nodes display a pair of numbers whose product equals 72. Since both factors, 8 and 9, are composite numbers, we keep decomposing them into products of two factors by issuing two new branches emanating from each node. Once prime factors are obtained, their nodes are declared as **leaves**—no branch emanates from them downward. At the end of the process, the leaves of the tree comprise the prime factors of 72.

Other factorizations of 72 are possible. For example, consider this tree diagram:

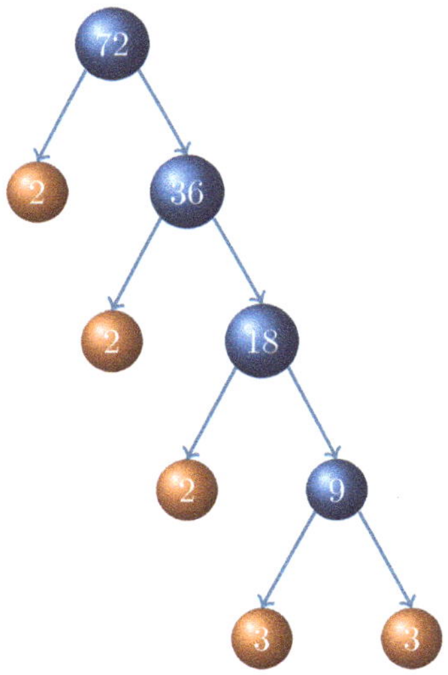

Yet, as guaranteed by the second part of the fundamental theorem of arithmetic, no matter how we grow the tree, we always end up with the same set of leaves. In other words, no matter how we decompose 72, we always obtain the same collection of prime factors, 2, 2, 2 , 3 and 3.

This last tree diagram for decomposing 72 is based on a systematic approach, which we exemplify by decomposing a larger number, 2660:

(a) Is 2660 divisible by the smallest prime number, 2? The answer is positive, hence we decompose 2660 into

$$2660 = 2 \times 1330.$$

Please note that doing so requires knowing whether a number is divisible by 2—a skill children acquire already in first grade—and requires being able to divide a number by 2.

(b) We already know that the first factor, 2, is prime, so what is left over is to decompose the second factor, 1330. Is 1330 divisible by the smallest prime number, 2? The answer is positive, hence we decompose it into

$$1330 = 2 \times 665.$$

(c) We now have to decompose the factor 665. Is it divisible by the smallest prime number, 2? Since the answer is negative, we proceed to check whether it is divisible by the next smallest prime number, 3. Here too the answer is negative, so we proceed to check whether it is divisible by the next smallest prime number, 5. This time the answer is positive, so we decompose 665 into

$$665 = 5 \times 133.$$

(d) Next, we decompose 133. Since its multiple 665 was found not to be divisible by neither 2 nor 3, we deduce, by the transitivity of the divisor-of property (p. 87) that 133 is not divisible by neither 2 nor 3 as well. Is 133 divisible by the next smallest prime number, 5? No. Is it divisible by the next smallest prime number, 7. Yes. so we decompose 133 into

$$133 = 7 \times 19.$$

(e) Next, we have to decompose the factor 19, which turns out to be a prime number.

(f) To conclude, we have decomposed 2660 into a product of prime numbers,

$$2660 = 2 \times 2 \times 5 \times 7 \times 19.$$

23.2.3 Why is 1 not a prime number?

We defined prime numbers as natural numbers having *exactly* two divisors. Another common definition is that a natural number n is prime if its only divisors are 1 and n. If we stick to this definition, then 1 should be considered as a prime number, as it is divisible by 1, it is divisible by itself, and it is not divisible by any other number. Yet, 1 is not considered a prime number.

The reason for excluding 1 from the set of primes is that: prime numbers are considered as the building blocks of all natural numbers, with the multiplication operation being the glue binding them. Since 1 is the identity element of multiplication (Volume 1, p. 87), multiplying by 1 does not "build a new number", hence 1 does not qualify as a building block for making other numbers.

23.2.4 Prime number factorization: The fingerprints of natural numbers

Since every natural number (other than 1) is either a prime, or it has only one representation as a product of prime numbers, one can view the primes composing a number as its "fingerprints", i.e., a characteristic that exclusively identifies that number.

For example, the collection of prime numbers that compose 72 constitutes its fingerprint, and so does the collection of prime numbers that compose 2660. Finally, since 3 is a prime number, its fingerprint comprises its unique prime factor, 3:

The collection of prime factors composing the natural number n will be marked by $\mathcal{P}(n)$, for example:

$$\mathcal{P}(72) = \{2,2,2,3,3\}$$
$$\mathcal{P}(2660) = \{2,2,5,7,19\}$$
$$\mathcal{P}(3) = \{3\}.$$

To state this in words, the collection of primes composing 72 is 2, 2, 2, 3 and 3, the collection of primes composing 2660 is 2, 2, 5, 7 and 19, and the collection of primes composing 3 is just the number 3 itself.

Another way to represent a natural number's fingerprint is to enumerate the **multiplicity** (i.e., the number of appearances) of each prime number in its prime number factorization. Take for example the number 72. Its prime number factorization can be represented by the following table:

Prime factor	Multiplicity
2	3
3	2

Similarly, the prime number factorization of 2660 can be represented by the following table:

Prime factor	Multiplicity
2	2
5	1
7	1
19	1

23.3 Enumerating the divisors of a given number

Now that we know how to decompose natural numbers into products of prime factors, we return to a subject considered earlier: finding efficiently all the divisors of a given natural number.

Consider, for example, the natural number 24, whose divisors were listed on p. 88. The first stage in finding all its divisors is to find its prime number factorization—its fingerprint:

$$\mathcal{P}(24) = \{2, 2, 2, 3\}.$$

We then proceed as follows:

(a) Every prime factor of 24 is a divisor. Please note that although a prime number may appear more than once in a number's factorizaton, we count it only once when enumerating the number's divisors. Thus, we have found that 24 is divisible by 2 and 3.

(b) Every product of prime factors of 24 is also a divisor of 24. For example, 2×2 is a divisor of 24 because

$$24 = (2 \times 2) \times (2 \times 3).$$

Thus, we find more divisors of 24 by forming all possible products of its prime factors,

$$\begin{aligned} 2 \times 2 &= 4 \\ 2 \times 3 &= 6 \\ 2 \times 2 \times 2 &= 8 \\ 2 \times 2 \times 3 &= 12 \\ 2 \times 2 \times 2 \times 3 &= 24. \end{aligned}$$

(c) Finally, 1 is a divisor of 24, since it is a divisor of every natural number.

(d) Since every divisor of 24 is either 1, or a a prime factor of 24 or a product of prime factors of 24, there are no other divisors than the ones enumerated above: 1, 2, 3, 4, 6, 8, 12 and 24.

Actually, this way of finding all the divisors of 24 involves finding all the subsets of its prime factors. Indeed, the prime number factorization of all the divisors of 24 is

$$\begin{aligned} \mathcal{P}(1) &= \{\} \\ \mathcal{P}(2) &= \{2\} \\ \mathcal{P}(3) &= \{3\} \\ \mathcal{P}(4) &= \{2, 2\} \\ \mathcal{P}(6) &= \{2, 3\} \\ \mathcal{P}(8) &= \{2, 2, 2\} \\ \mathcal{P}(12) &= \{2, 2, 3\} \\ \mathcal{P}(24) &= \{2, 2, 2, 3\}, \end{aligned}$$

where $\{\}$ marks the **empty set**—there are no prime numbers compositing 1. In other words, the collection of prime factors of 24 comprises the building blocks of all its divisors.

23.4 Prime numbers and rectangular arrays

We have seen on several occasions that a very convenient illustration of multiplication is by a **rectangular array**. Since the **primality** of a number (the property of being prime) is intimately related to the multiplication operation, it is not surprising that primality can be characterized using a rectangular array metaphor.

Consider again the number 24. Its smallest divisor is 1, and the divisor complementing it is 24, i.e., $24 = 1 \times 24$. This fact can be illustrated by arranging 24 items in a single row comprising 24 items, i.e., in a **linear array**:

The next smallest divisor of 24 is 2, and the divisor complementing it is 12, i.e., $24 = 2 \times 12$. This fact can be illustrated by arranging 24 items in a rectangular array of 2 rows comprising 12 items each:

The next smallest divisor of 24 is 3, and the divisor complementing it is 8, i.e., $24 = 3 \times 8$. This fact can be illustrated by arranging 24 items in a rectangular array of 3 rows comprising 8 items each:

The next smallest divisor of 24 is 4, and the divisor complementing it is 6, i.e., $24 = 4 \times 6$. This fact can be illustrated by arranging 24 items in a rectangular array of 4 rows comprising 6 items each:

With this we have exhausted all the possible arrangements of 24 items in linear or rectangular arrays (up to rotations by 90°, which corresponds to interchanging the order of the factors).

We can now characterize prime numbers using rectangular arrays:

> A natural number n greater than 1 is a prime number if n items can not by arranged in a rectangular array, but only in a linear array.

Activity: Rectangular arrays

Rectangular arrays can help children develop an intuitive understanding of what it takes to be a prime number. The following activity can precede a more formal study of prime numbers:

(a) Present children different ways of arranging a set of items in a rectangular array.
(b) Ask the children to create rectangular arrays with 2 items, with 3 items, and so on up to 30 items.
(c) Ask the children to write down in which cases they did not succeed in assembling rectangular arrays.

23.5 More on prime numbers

23.5.1 The sieve of Eratosthenes

Since humans discovered the existence of prime numbers, the race was on to seek systematic ways to find them. The first known algorithm achieving this goal is attributed to Eratosthenes, a Greek mathematician who lived in the 3rd century BCE, and served as a librarian of the world famous Alexandria library. His method for listing the prime numbers was as follows:

First, write all the natural numbers greater than 1 in a row.

2 3 4 5 6 7 8 9 10 11 12 13 14 15 16 17 18 19 20 21 22 23 24 25 26 27 ⋯

Obviously, one cannot write an infinite sequence of numbers. This is not important. What matters is that, *in principle*, every natural number can *eventually* be listed.

The first number in the sequence, 2, must be prime because it cannot be divisible by any smaller number other than 1. Mark it, and eliminate all of its multiples by deleting every second number on the list. These numbers are guaranteed not to be primes because they are divisible by 2.

(2) 3 ~~4~~ 5 ~~6~~ 7 ~~8~~ 9 ~~10~~ 11 ~~12~~ 13 ~~14~~ 15 ~~16~~ 17 ~~18~~ 19 ~~20~~ 21 ~~22~~ 23 ~~24~~ 25 ~~26~~ 27 ⋯

Next, ignore the numbers that were deleted, and locate the next undeleted number in the list succeeding 2. This number happens to be 3. It must be a prime number, otherwise it would have been a multiple of 2, and it would have been deleted. Mark it, and eliminate all its multiples by deleting every third number in the list. Note that some of these numbers were already deleted, being also multiples of 2.

(2) (3) ~~4~~ 5 ~~6~~ 7 ~~8~~ ~~9~~ ~~10~~ 11 ~~12~~ 13 ~~14~~ ~~15~~ ~~16~~ 17 ~~18~~ 19 ~~20~~ ~~21~~ ~~22~~ 23 ~~24~~ 25 ~~26~~ ~~27~~ ⋯

Next, keep ignoring the numbers that were deleted, and locate the next undeleted number succeeding 3. This number happens to be 5. It must be a prime number, otherwise it must have been a multiple of either 2 or 3, and it would have been deleted. Mark it and eliminate all its multiples by deleting every fifth number in the list.

(2) (3) ~~4~~ (5) ~~6~~ 7 ~~8~~ ~~9~~ ~~10~~ 11 ~~12~~ 13 ~~14~~ ~~15~~ ~~16~~ 17 ~~18~~ 19 ~~20~~ ~~21~~ ~~22~~ 23 ~~24~~ ~~25~~ ~~26~~ ~~27~~ ⋯

Keep proceeding this way. At each phase, it is guaranteed that the next undeleted number in the list is the succeeding prime.

This systematic way of retaining the prime numbers and eliminating the composite numbers is visioned as a **sieve**, the metaphor being the separation of small grains (the composite numbers) from the large ones (the prime numbers).

23.5.2 There are infinitely many prime numbers

When we examine the sequence of prime numbers, we discover that succeeding primes tend to be, on average, more and more separated. The natural question to ask is: does the sequence of prime numbers ever terminate? Are there finitely many prime numbers, or is there an infinite number of prime numbers?

About 2500 years ago, the Greek mathematician Euclid proved that the sequence of prime numbers never terminates—there is an infinite number of prime numbers. His proof is both sophisticated and astonishing in its simplicity. Therefore, it provides an attractive opportunity to show off a great piece of mathematics to school children.

Euclid proved that there is an infinite number of prime numbers by using a strategy known as **proof by contradiction**. He conducted a thought experiment, assuming that there are only finitely many prime numbers, and showed that this leads to a contradiction. Hence, he concluded, there must be an infinite number of prime numbers.

Let's follow Euclid's step and prove:

Claim 23.2. *There is an infinite number of prime numbers.*

Euclid's proof

(1) Assume that there were only finitely many prime numbers. That is, there exists

a finite set, let's call if P (for *primes*), containing *all* the prime numbers. This assumption also implies that there is one prime number that is greater than all other prime numbers.

(2) Take all the numbers in the set P, calculate their product and denote this product by n. This is a *thought experiment*, so you don't really need to calculate anything. What matters is that since P contains finitely many numbers, it would be possible, *in principle*, to multiply all the numbers in P.

(3) The number n obtained by multiplying all the number in P has a very special property. It is a multiple of *all* prime numbers.

(4) Next, consider the number $n+1$. It cannot be a prime number because it is greater than all the numbers in P. Moreover, since n is divisible by all the prime numbers, $n+1$ is not divisible by any prime number; if we divide $n+1$ by a prime number we always obtain a remainder of 1.

(5) Thus, $n+1$ is neither prime, nor it is a product of prime numbers. This contradicts the fundamental theorem of arithmetics, which states that every natural number is either prime or a product of prime numbers.

Let's demonstrate Euclid's proof with an example. Assume someone claimed that there were only three prime numbers, 2, 3 and 5. You can prove it incorrect by showing that 7 is also a prime number, but you could also prove it incorrect without calculating anything. All you need to do is to argue that adding 1 to the product of "all" the prime numbers,

$$2 \times 3 \times 5 + 1$$

yields a number not divisible by neither 2, 3 or 5. Hence either it is a prime number (thus contradicting their claim), or it is a composite number not divisible by any prime number, which is impossible.

About proofs by contradiction

Proofs by contradiction are commonly used in mathematics. Behind them lies the following idea: We want to determine that a certain proposition is true (in Euclid's case, the proposition is that there is an infinite number of prime numbers). There are two possibilities: Either the proposition is true or it is false. Next we conduct a thought experiment, assuming that the proposition is actually false, and perform a sequence of logical deductions based on this assumption. If, as a result, we reach a conclusion which we know to be incorrect (in mathematicians' jargon, reach a contradiction), then we must rule out the possibility that the proposition is false, hence it is true.

23.5.3 Large prime numbers

Since the time of Euclid, we have known that there is an infinite number of prime numbers. This doesn't mean, however, that we actually know how to find as many prime numbers as we want. Moreover, if you construct a huge number by randomly writing a very long sequence of digits, checking whether this number is prime or composite might not be feasible, even when aided by powerful super-computers.

Over the years, there has been a continual ongoing race to find the largest prime number known to date. Naturally, nowadays the task of finding very large prime numbers makes extensive use of computers. Every several years, newspapers announce the discovery of a new prime number, larger than the precedent record holder. As of the writing of these lines, the largest prime number known to us comprises 22,338,618 digits. Just writing it down would require about 8000 pages.

23.5.4 The density of prime numbers

Consider all the prime numbers up to 10: 2, 3, 5, and 7. They constitute 40% of all natural numbers in this range. Next, examine all the prime numbers up to 100. There are 25 such numbers, constituting 25% of all natural numbers in this range. The following table displays the percentage of prime numbers among all natural numbers up to 1,000, 10,000, etc:

Range	Percentage of primes
1–10	40%
1–100	25%
1–1,000	16.8%
1–10,000	12.2%
1–100,000	9.5%
1–1,000,000	7.8%
1–10,000,000	6.6%
1–100,000,000	5.7%

This table reveals two interesting facts:

(1) The prime numbers are abundant. Up to 100 million, more than one number out of 20 is prime.
(2) The density of prime numbers decreases continually as the range of numbers increases.

Mathematicians have spent considerable efforts over centuries to understand how fast the density of the primes decreases at the number range increases. At the end of the 18th century, German mathematician Carl Friedrich Gauss conjectured that the number of prime numbers up to n equals, with very high accuracy, to n divided by the natural logarithm of n. This conjecture—now a theorem—was proved in 1896 concurrently by the Belgian mathematician Charles Jean de la Vallée-Poussin (1866–1962) and the French mathematician Jacques Salomon Hadamard (1865–1963).

23.5.5 Twin prime numbers

The smallest prime number, 2, is the only prime number that is even. From this follows that from 3 onward, the difference between two consecutive primes is at least 2. Examining the list of prime numbers up to 100, we see that there are eight pairs of prime numbers that differ by 2. Listing them all,

$$3,5 \quad 5,7 \quad 11,13 \quad 17,19 \quad 29,31 \quad 41,43 \quad 59,61 \quad 71,73.$$

Pairs of prime numbers differing by 2 are called **twin prime numbers**.

Mathematicians have been asking themselves whether there is an infinite number of pairs of twin prime numbers, or conversely, whether there exists a number beyond which there are no more twin primes. Note that the fact that there is an infinite number of primes does not guarantee the existence of an infinite number of twin primes. Aided by computers, mathematicians have strived to find pairs of twin primes of ever increasing magnitude. The largest pair of twin primes discovered to date contains $388,342$ digits.

Nevertheless, it is important to stress: The fact that new twin primes are continually discovered does not guarantee that there is an infinite number of them. Until today, the infinitude of twin primes has remained an open question. Mathematician all over the world are still trying to prove or disprove the existence of an infinite number of twin primes.

23.5.6 Goldbach's conjecture

A conjecture is a proposition believed to be true, yet, so far, no one has managed to prove or disprove it. In 1742, the German mathematician, Christian Goldbach (1690–1764) made a conjecture now bearing his name:

> Every even number greater than 2 can be expressed as a sum of two prime numbers.

Let's check whether Goldbach's conjecture holds for even numbers in the range 4–18:

$$4 = 2 + 2 \qquad 6 = 3 + 3 \qquad 8 = 3 + 5 \qquad 10 = 3 + 7$$
$$12 = 5 + 7 \qquad 14 = 3 + 11 \qquad 16 = 3 + 13 \qquad 18 = 7 + 11.$$

Computer experiments revealed that Goldbach's conjecture holds true up to huge numbers—a billion billions. This does not rule out, however, the possibility that there exists some larger even number that cannot be written as a sum of two primes.

One of the beauties of this conjecture, is that every elementary school student can understand its content. Yet, for over 250 years, some of the world's greatest minds have failed to either prove or disprove Goldbach's conjecture.

Mathematical problems

Problem 23.1. Consider the natural number 5940.

(a) Represent it as a product of prime factors.
(b) Find all its divisors.

Problem 23.2. There are 29 beans in the plate. In how many ways can they be divided into equal-sized sets?

Problem 23.3. Are there two prime numbers whose sum is also a prime? How does this question relate to a celebrated conjecture?

Problem 23.4. In order to test whether 1379 is a prime number, one has to divide it by different numbers and check whether or not there is a remainder. What numbers *must* you divide 1379 by if you want to be certain that it is prime? The goal is to minimize the amount of computations.

Problem 23.5. Find two prime numbers between 100 and 200, in which both of their ones and tens digits are primes. Additionally, if the hundreds digit is deleted, the resulting two-digit number is also a prime number.

Problem 23.6. Check whether Goldbach's conjecture holds for all even numbers up to 200.

Problem 23.7. There once was an infinitely long street, with an infinite number of street lights numbered 1, 2, etc. At the beginning of our story all the lamps were turned off.

On Day 1, the lamplighter lit all the street lights:

On Day 2, he turned off every other light, starting from that labeled 2:

On Day 3, he changed the state of every third light. If that light was off, he turned it on, and if that light was on, he turned it off.

On Day 4, he changed the state of every fourth light, and so he carried on day after day. On Day n he changed the state of all the streetlights numbered by a multiple of n.

Which streetlights will remain lit forever after he has visited them for the last time?

Problem 23.8. Throughout history, mathematicians have attempted to develop formulas generating prime numbers. The most famous of them was invented by Swiss mathematician Leonhard Euler (1707–1783). For every n, consider the number

$$n^2 + n + 41.$$

If, for example, we substitute $n = 1$, we obtain 43, which is a prime number. If we substitute $n = 2$, we obtain 47, which is also a prime number.

(a) Substitute $n = 3$, $n = 4$, and $n = 5$, and check whether a prime number is obtained.
(b) Without calculating anything, does substituting $n = 41$ yield a prime number? Explain your answer.

Pedagogical problems

Problem 23.9. Gili claims that a prime number is one that does not appear in the multiplication table.

(a) Explain what Gili means.
(b) Is she right?

Problem 23.10. Are there two composite numbers whose sum is a prime?

Alon, in answering this question, claimed that there are no such numbers. His argument was that a sum can always be broken down back into its addends.

Tal also claimed that there are no such numbers. His argument was that 4 and 6, for example, are composite numbers and their sum is 10, which is also a composite number.

(a) What is your opinion of Alon's and Tal's answers?
(b) Formulate precise replies to Alon and Tal.

Problem 23.11. In the context of prime number factorization, Teacher Cam recorded in class the following statements:

Mai said: If x is larger than y, then the largest prime divisor of x must be larger than the largest prime divisor of y.
Matan said: If x is larger than y, then the smallest prime divisor of x must be larger than the smallest prime divisor of y.
Sivan said: If x is larger than y, then x must have more prime factors than y.

Formulate a reply to each child.

Chapter 24

Common Multiples and Common Divisors

In Chapter 23 we encountered the notions of **multiples** and **divisors**. In particular, we learned about prime numbers as the building blocks of all the natural numbers. In this chapter we will elaborate on these topics, and develop tools necessary later on in the context of fractions.

24.1 Least common multiple

24.1.1 Common multiples

We start with the following definition:

Definition 24.1. Let a and b be natural numbers. A natural number n is called a **common multiple** of a and b if it is a multiple of both numbers. Using the **divisor-of** notation introduced on p. 86, n is a common multiple of a and b if

$$a \lhd n \qquad \text{and} \qquad b \lhd n.$$

Consider, for example, the numbers 36 and 48. *Do they have a common multiple?*

The answer is positive: their product, 36×48, is clearly a multiple of both 36 and 48. We don't even need to evaluate this product (which happens to be 1728) to know it is a common multiple.

Having determined the existence of at least one common multiple of 36 and 48, we may ask: *Do* 36 *and* 48 *have other common multiples?*

The answer is again positive. Every multiple of 1728 is automatically a multiple of both 36 and 48, for example,

$$2 \times 1728 = 3456$$

is also a common multiple of 36 and 48. This fact stems from the transitivity of the multiple-of relation (p. 87). Indeed,

$$\text{If} \quad 36 \lhd 1728 \quad \text{and} \quad 1728 \lhd 3456 \quad \text{then} \quad 36 \lhd 3456,$$

and

$$\text{If} \quad 48 \lhd 1728 \quad \text{and} \quad 1728 \lhd 3456 \quad \text{then} \quad 48 \lhd 3456.$$

That is, 3456 is a common multiple of both 36 and 48. Since 1728 has infinitely many multiples, it follows at once that 36 and 48 have infinitely many common multiples.

Having realized that 36 and 48 have infinitely many common multiples, we finally ask: *What is the least common multiple of* 36 *and* 48*?*

This question turns out to be harder than the previous ones. One of the aims of this chapter is to learn systematic and efficient methods for evaluating the least common multiple of any two numbers. The skill of finding common multiples becomes important later on, when dealing with fractions. It is a prerequisite for finding a common denominator of two fractions, and therefore, a prerequisite for comparing, adding and subtracting fractions with unlike denominators (see Chapter 30 below). Finding the least common multiple is useful in order to obtain the simplest possible fractions with like denominators. Finally, understanding properties of common multiples is a subject interesting in its own, offering an excellent opportunity to expand our mastery in numbers and operations.

Notation: given two integers a and b, we denote their least common multiple by

$$\text{LCM}(a, b).$$

24.1.2 Evaluation of the least common multiple

24.1.2.1 Listing of multiples

The simplest method, yet not the most efficient one, for finding the least common multiple of two numbers is to list their multiples one-by-one until a common multiple is found. Since every number has infinitely many multiples, we need to know "when to stop". We will demonstrate this strategy through several examples.

Example 1: Find the least common multiple of 5 and 7.

We proceed as follows:

(a) Evaluate the product of the two numbers, $5 \times 7 = 35$. We know that this product is a common multiple. Therefore, we can be certain that the least common multiple of 5 and 7 is not greater than 35.

(b) List the multiples of 5 (that is, skip-count by 5) and list the multiples of 7 (that is, skip-count by 7). In each sequence, stop once reaching 35:

$$5, 10, 15, 20, 25, 30, 35$$
$$7, 14, 21, 28, 35.$$

(c) Examine both sequences, and look for the smallest number appearing in both sequences. In this case, 35 is the only number appearing in both lists, thus, we deduce that 35 is the least common multiple of 5 and 7. In our notation,

$$\text{LCM}(5, 7) = 35.$$

This approach is systematic and, more importantly, guaranteed to yield the correct answer after a number of steps known in advance. Its shortcoming becomes clear when dealing with larger numbers. This can be demonstrated using the example with which we opened this chapter:

Example 2: Find the least common multiple of 36 and 48.

The product of 36 and 48 is 1728. Listing all the multiples of 36 and 48 up to 1728 requires listing 48 multiples of 36 and 36 multiples of 48. Quite a daunting—and unnecessary—task.

Instead, we may use the following modification of the above algorithm: generate the two lists of multiples simultaneously, progressing at each step the sequence whose last element is the smallest, until a common multiple has been reached.

Specifically, since 36 is less than 48, we calculate first the next multiple of 36:

$$36, 72$$

$$48.$$

Since 48 is less than 72, we evaluate the next multiple of 48

$$36, 72$$

$$48, 96.$$

Since 72 is less than 96, we evaluate the next multiple of 36

$$36, 72, 108$$

$$48, 96.$$

Since 96 is less than 108, we evaluate the next multiple of 48

$$36, 72, 108$$

$$48, 96, 144.$$

Since 108 is less than 144, we evaluate the next multiple of 36

$$36, 72, 108, 144$$

$$48, 96, 144.$$

We have thus obtained a common multiple, 144, which by the very construction is the least common multiple,

$$\mathrm{LCM}(36, 48) = 144.$$

Indeed, if there existed a smaller common multiple, this process would have stopped earlier.

This refined method is more efficient than the original one, but may still require a great deal of unnecessary computations, as shown by the next example:

Example 3: Find the least common multiple of 36 and 37.

If we adopted the same strategy as in Example 2, we would have to list all the multiples of 36 and 37 up to their product, $36 \times 37 = 1332$, which happens to be their least common multiple. This last example motivates us to look for more efficient methods for finding least common multiples.

24.1.2.2 Using prime number factorization

As argued and demonstrated in Chapter 23, many properties of natural numbers can be inferred by considering their fingerprints—their prime factors.

Consider again Example 2 above: Find the least common multiple of 36 and 48. Using prime number factorization, we solve the problem in three stages:

(a) Find the prime number factorization of both numbers (p. 95):

$$\mathcal{P}(36) = \{2, 2, 3, 3\}$$
$$\mathcal{P}(48) = \{2, 2, 2, 2, 3\}.$$

(b) Every multiple of 36 is, by definition, divisible by 36. Hence, the collection of prime numbers compositing it must contain all the prime factors composing 36. Likewise, every multiple of 48 is divisible by 48. Hence, the collection of prime numbers compositing it must contain all the factors composing 48. Thus, we may conclude the following facts:

- Every multiple of 36 contains the factor 2 at least twice.
- Every multiple of 48 contains the factor 2 at least four times.
- Every multiple of 36 contains the factor 3 at least twice.
- Every multiple of 48 contains the factor 3 at least once.

(c) Every common multiple of 36 and 48 must meet all four conditions listed above. It must contain the factor 2 at least four times and the factor 3 at least twice. The least common multiple will contain *only* these factors, hence it will be equal to

$$\text{LCM}(36, 48) = 2 \times 2 \times 2 \times 2 \times 3 \times 3 = 144.$$

This technique for finding the least common multiple of two numbers can be organized in the form of a table. The first entry in each row exhibits a prime factor composing either 36 or 48. The second entry displays the number of occurrences of that factor in $\mathcal{P}(36)$. The third entry displays the number of occurrences of that factor in $\mathcal{P}(48)$. The fourth entry is the largest between the second and third entry: it is the number of times that factor occurs in the prime number factorization of their least common multiple.

Prime factor	# times in $\mathcal{P}(36)$	# times in $\mathcal{P}(48)$	# times in $\mathcal{P}(\text{LCM}(36, 48))$
2	2	4	4
3	2	1	2

Another illustration of the same solution strategy uses **Venn diagrams**. The collection of primes composing 36 and 48 are represented as two sets in the following manner:

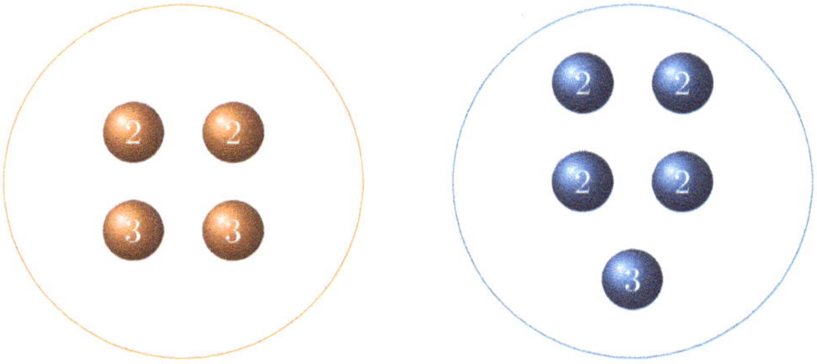

$\mathcal{P}(36)$ $\mathcal{P}(48)$

The prime factors composing the least common multiple of 36 and 48 is the **union** of their prime factors. As always, when forming unions, an element that appears in both collections is accounted for only once, and placed in the area where both collections intersect. Thus, the union of the collections of prime factors composing 36 and 48 can be represented as follows:

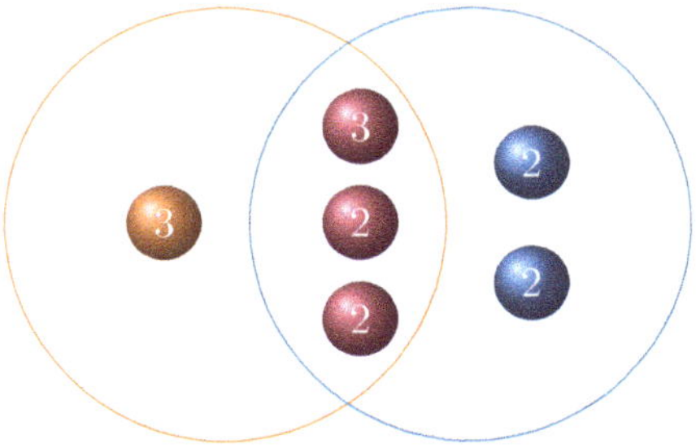

$\mathcal{P}(\mathrm{LCM}(36, 48))$

Every multiple of 36 must contain all the prime factors circled in orange and every multiple of 48 must contain all the prime factors circled in blue. A common multiple of both numbers must contain the combination of the two collections of factors, and their least common multiple contains *only* those factors.

24.2 Greatest common divisor

24.2.1 Common divisors

We start with the following definition:

Definition 24.2. Let a and b be natural numbers. A natural number n is called a **common divisor** of a and b if it is a divisor of both numbers. Using the divisor-of notation introduced on p. 86, n is a common divisor of a and b if

$$n \lhd a \qquad \text{and} \qquad n \lhd b.$$

Consider the natural numbers 72 and 96. *Do they have a common divisor?*

The answer is positive. Every two numbers have at least one common divisor—1 (see p. 86).

Having determined the existence of at least one common divisor, we next ask: *Do* 72 *and* 96 *have a common divisor other than* 1*?*

Once again the answer is positive. 2 is also a common divisor of both numbers. We further ask: *Do* 72 *and* 96 *have infinitely many common divisors?*

This time the answer is negative. The greatest divisor of 72 is 72 and the greatest divisor of 96 is 96. It follows that a common divisor of any two numbers cannot be greater than the least of those number—in this case, 72.

Finally, we ask: *What is the greatest common divisor of* 72 *and* 96*?*

The answer to this question is more involved, and is the focus of the present section. The notion of greatest common divisor has many ramifications in mathematics. In the elementary school curriculum, common divisors are used in **fraction reduction**, also known as **fraction simplification** (see Chapter 28). Moreover, to bring a fraction into irreducible form, one has to divide both numerator and denominator by their greatest common divisor.

Notation: given two integers a and b, we denote their greatest common divisor by

$$\text{GCD}(a, b).$$

24.2.2 Evaluation of the greatest common divisor

24.2.2.1 Listing all the divisors

The simplest method, yet not the most efficient one, for finding the greatest common divisor of two numbers is to list all their divisors, and then inspect the two lists to find the greatest number appearing in both.

Take for example the numbers 72 and 96. All their divisors are listed below:

Number	Divisors
72	1,2,3,4,6,8,9,12,18,24,36,72
96	1,2,3,4,6,8,12,16,24,32,48,96

Examining the two lists shows that there are eight common divisors, the largest of which being 24, i.e.,

$$\text{GCD}(72, 96) = 24.$$

The drawback of this method is its complexity when dealing with larger numbers. As in the case of the least common multiple, more efficient techniques can be derived.

24.2.2.2 Using prime number factorization

Consider again the numbers 72 and 94. Their prime number factorization is:

$$\mathcal{P}(72) = \{2,2,2,3,3\}$$
$$\mathcal{P}(96) = \{2,2,2,2,2,3\}.$$

Remember the important fact noted on p. 96: All the divisors of a number are composed from the prime factors of that number. Thus,

- All the divisors of 72 are composed from the factors 2, 2, 2, 3 and 3.
- All the divisors of 96 are composed from the factors 2, 2, 2, 2, 2 and 3.

Thus, we conclude that all the common divisors of 72 and 96 are composed from prime factors joint to both numbers.

Specifically,

- Any divisor of 72 contains the factor 2 at most three times.
- Any divisor of 96 contains the factor 2 at most five times.
- Any divisor of 72 contains the factor 3 at most twice.
- Any divisor of 96 contains the factor 3 at most once,

from which we conclude that

- Any common divisor of 72 and 96 contains the factor 2 at most three times.
- Any common divisor of 72 and 96 contains the factor 3 at most once.

The greatest common divisor is that common divisor containing the *maximum* number of prime factors allowed by the above constraints, i.e.,

$$\text{GCD}(72,96) = 2 \times 2 \times 2 \times 3 = 24.$$

This means for finding the greatest common divisor of two numbers can be organized in the form of a table. The first entry in each row exhibits a prime factor composing either 72 or 96. The second entry displays the number of occurrences of that factor in $\mathcal{P}(72)$. The third entry displays the number of occurrences of that factor in $\mathcal{P}(96)$. The fourth entry is the least between the second and third entry: it is the number of times that factor occurs in the prime number factorization of their greatest common divisor.

Prime factor	# times in $\mathcal{P}(72)$	# times in $\mathcal{P}(96)$	# times $\mathcal{P}(\text{GCD}(72,96))$
2	3	5	3
3	2	1	1

The same strategy can be illustrated using Venn diagrams. The collections of primes composing 72 and 96 are represented as two sets:

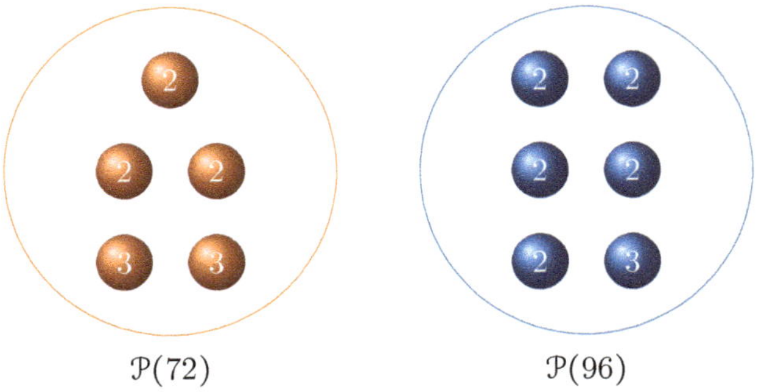

The prime factors composing their greatest common divisor is the **intersection** of their prime factors, as depicted below:

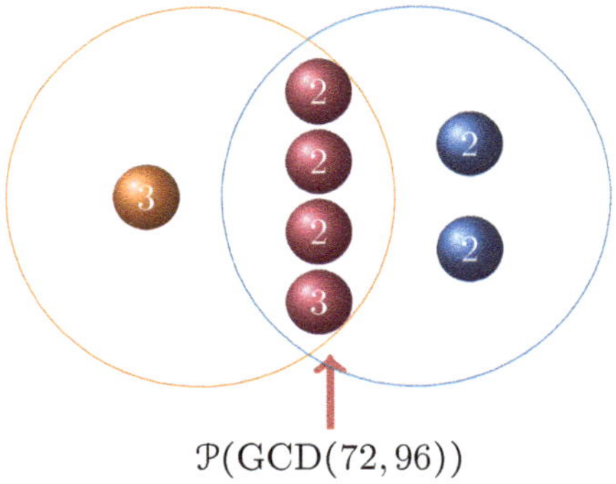

Every divisor of 72 is composed of prime factors circled in orange and every divisor of 96 is composed of prime factors circled in blue. A common divisor must be composed of prime factors enclosed in both circles. The greatest common divisor is composed of *all* prime factors enclosed in both circles.

24.2.2.3 The Euclidean Algorithm

An even more efficient algorithm for finding the greatest common divisor of two numbers is attributed to Euclid. The idea behind it is based on the following observation.

If n is a divisor of a and b, then it is also a divisor of their difference.

In formal notation,

$$\text{If } n \lhd a \text{ and } n \lhd b, \text{ then } n \lhd (b - a).$$

For example, the difference of any two even numbers is even, which is the same as saying that if 2 is a divisor of two numbers, then it is also a divisor of their difference.

How can this fact be exploited to find greatest common divisors?

Take for example the numbers 72 and 104.

(a) Any common divisor of 72 and 104 is also a divisor of their difference $104-72=32$.
(b) Concentrate then on the two smallest numbers, 72 and 32. Any common divisor of 72 and 32 is also a divisor of their difference $72-32=40$.
(c) Now take 32 and 40: any common divisor of 32 and 40 is also a divisor of their difference $40-32=8$.
(d) Any common divisor of 32 and 8 is also a divisor of their difference $32-8=24$.
(e) Any common divisor of 24 and 8 is also a divisor of their difference $24-8=16$.
(f) Any common divisor of 24 and 8 is also a divisor of their difference $16-8=8$.

If we rewind this chain of deduction, we find that any common divisor of the two original numbers, 72 and 104, must also be a divisor of 8. Since the largest divisor of 8 is 8 itself, it is the largest common divisor of 72 and 104.

The Euclidean algorithm has a very appealing geometric interpretation. Consider two rods, one 72 unit long (colored blue) and one 104 unit long (colored orange):

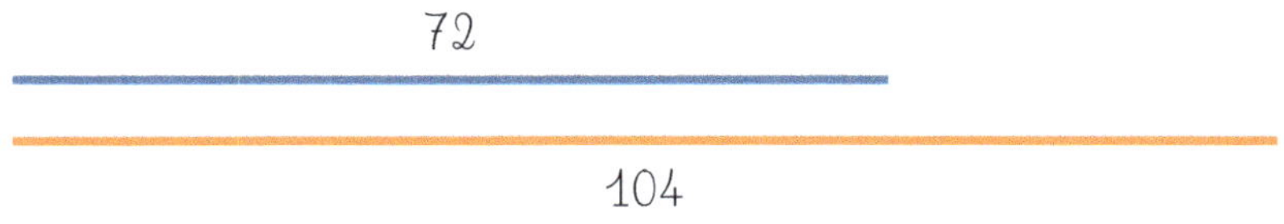

The greatest common divisor of 72 and 104 can be thought of as the integer length of the longest rod contained a whole number of times in both rods. Any such rod must also be contained a whole number of times in their difference, which is a 32 unit long rod (in green):

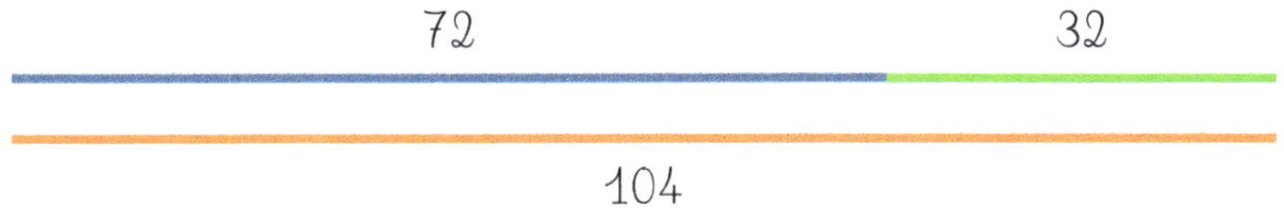

Thus, we have reduced the problem to finding the longest rod contained a whole number of times in a 72 unit long rod and a 32 unit long rod:

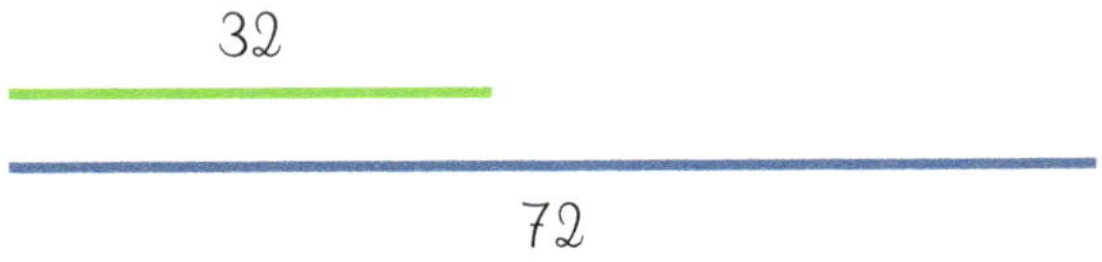

And so we proceed: Any rod contained a whole number of times in a 72 unit long rod and in a 32 unit long rod must also be contained a whole number of times in their difference, which is a 40 unit long rod (in purple):

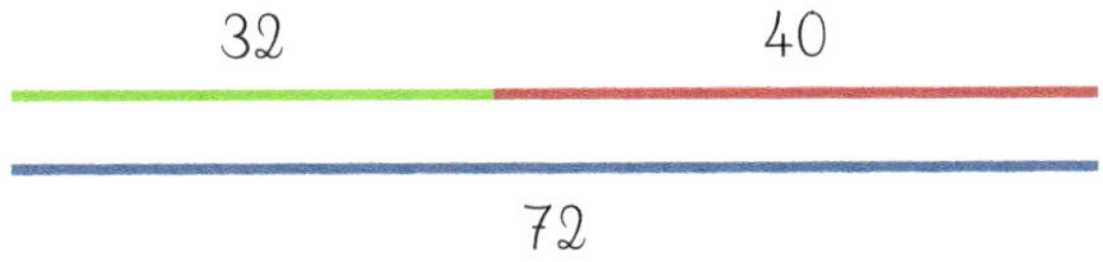

Thus, we have reduced the problem to finding the longest rod contained a whole number of times in a 32 unit long rod and a 40 unit long rod. The process continues until we remain with two rods, each 8 units long.

24.2.2.4 Rectangular array representation

Another interesting angle of view of the greatest common divisor uses rectangular arrays. This should not come as a surprise, as rectangular arrays are a standard representation whenever multiplication is involved. To illustrate the idea, consider the task of finding the greatest common divisor of 12 and 16.

We start by creating a rectangular array comprising 12 rows and 16 columns:

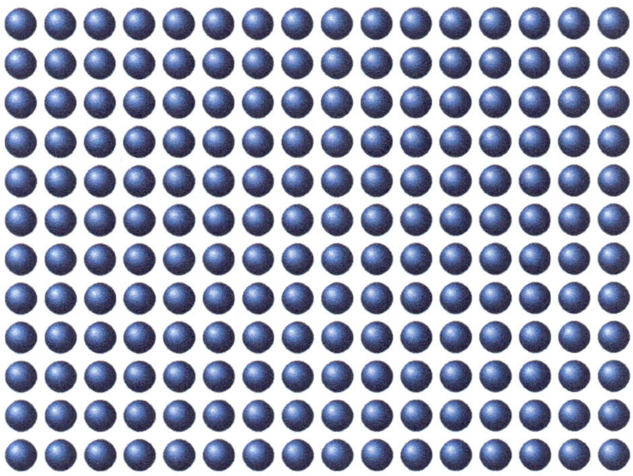

A number n is a divisor of 12 if we can partition 12 items into sets of n items each. It is also a divisor of 16 if we can partition 16 items into sets of n items each. If n is a common divisor of 12 and 16, we can partition this rectangular array into square arrays comprising n-by-n items.

For example, 2 is a common divisor of 12 and 16 because we can partition the rectangular array into squares of two-by-two, as follows:

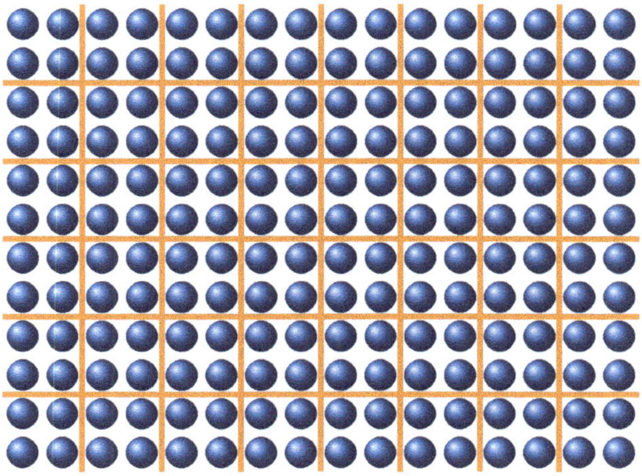

Likewise, 4 is the common divisor of 12 and 16 because we can partition the rectangular array into squares of four-by-four, as follows:

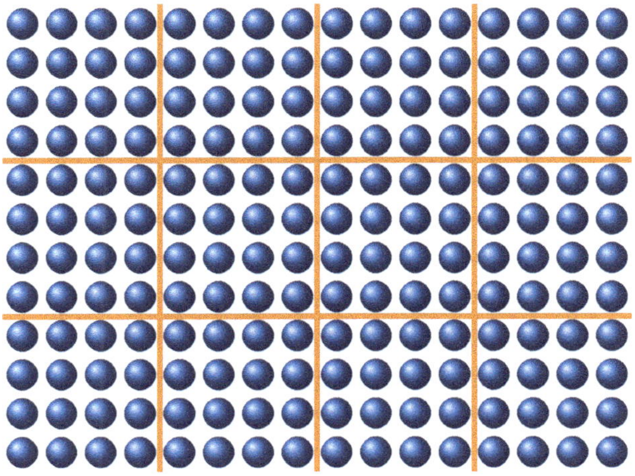

This viewpoint assigns common divisors another geometric meaning. If you generate a rectangular array from two given numbers, the greatest common divisor is the side length of the largest square with which you can tile the array. In the present example, it turns out that the largest square tiling the array has a 4 unit long side. Therefore,

$$\mathrm{GCD}(12, 16) = 4.$$

24.3 Relation between least common multiple and greatest common divisor

Given two numbers, their least common multiple and their greatest common divisor happen to satisfy an interesting relation. The easiest means to unravel this relation is, once again, through their prime number factorizations.

Take for example the pair of numbers 96 and 180. Their prime number factorizations are

$$\mathcal{P}(96) = \{2, 2, 2, 2, 2, 3\}$$
$$\mathcal{P}(180) = \{2, 2, 3, 3, 5\}.$$

As we have learned, the prime number factorization of their least common multiple is the union of their factors,

$$\mathcal{P}(\text{LCM}(96, 180)) = \{2, 2, 2, 2, 2, 3, 3, 5\}.$$

We also saw that the prime number factorization of their greatest common divisor is the intersection of their factors,

$$\mathcal{P}(\text{GCD}(96, 180)) = \{2, 2, 3\}.$$

Let's write these relations more explicitly:

$$96 = 2 \times 2 \times 2 \times 2 \times 2 \times 3$$
$$180 = 2 \times 2 \times 3 \times 3 \times 5$$
$$\text{LCM}(96, 180) = 2 \times 2 \times 2 \times 2 \times 2 \times 3 \times 3 \times 5$$
$$\text{GCD}(96, 180) = 2 \times 2 \times 3.$$

The prime factors on the left hand sides of the first two lines coincide with the prime factors on the left hand sides of the last two lines. This means that

$$\text{LCM}(72, 180) \times \text{GCD}(72, 180) = 72 \times 180.$$

More generally, for every two integers a and b,

$$\text{LCM}(a, b) \times \text{GCD}(a, b) = a \times b.$$

In particular, knowing the least common multiple of two numbers provides an easy way of finding their greatest common divisor, and vice versa.

Mathematical problems

Problem 24.1. What is the smallest number that is a common multiple of all the natural numbers between 1 and 10?

Problem 24.2. Is it true that if the product of two numbers is not a multiple of 7, then neither of the numbers is a multiple of 7? Justify your answer.

Problem 24.3. Is it true that if the product of two numbers is not a multiple of 9, then neither of the numbers is a multiple of 9? Justify your answer. Does this question differ from the previous one?

Problem 24.4. It is given that:

$$\mathcal{P}(144) = \{2, 2, 2, 2, 3, 3\}$$
$$\mathcal{P}(70) = \{2, 5, 7\}.$$

(a) What are the prime factors of the product 144×70?

(b) What is the least common multiple of 144 and 70?

(c) What is the greatest common divisor of 144 and 70?

Problem 24.5. The thickness of a math book is 8cm, whereas the thickness of a history book is 6cm. The bookstore has these two books stacked in two adjacent piles. How many books of each type can be stacked so that the piles are of equal height? Is there more than one answer? What is the smallest number of books needed for the two stacks to be of equal height?

Problem 24.6. What is the greatest common divisor of 21 and 16?

Problem 24.7. The car brakes require changing every other year; the oil filter requires changing every 3 years; the tires require changing every 4 years, and the belts require changing every 6 years. In how many years will everything need changing at the same time?

Problem 24.8. Complete the following table:

A	B	$A \times B$	LCM(A, B)	GCD(A, B)
6	9			
4	8			
7	14			
5	20			
12	16			
4	6			
3	5			
9	12			
10	30			
8	24			
9	36			
32	42			

Chapter 25

Divisibility Rules

A **divisibility rule** is an easy-to-apply criterion testing, without performing any division, whether numbers are divisible by a fixed given divisor. For example, a divisibility-by-3 rule is a simple criterion for testing whether numbers are divisible by 3.

In Chapter 6, we encountered a divisibility rule—the divisibility-by-2 rule, which determines whether numbers are odd or even. A whole number is divisible by 2 if and only if its ones digit is either 0, 2, 4, 6 or 8. Thus, for example, if we want to determine whether 2, 354, 014 is divisible by 2, there is no need to carry out a lengthy division operation. All one has to do is to examine the ones digit of that number, which is 4, to conclude that 2, 354, 014 is divisible by 2.

Hurdles:

Children apply divisibility rules blindly, without understanding their underlying principles.

Remedies:

Justify each divisibility rule; the underlying principles are within children's reach.

Opportunities:

Testing for divisibility and finding remainders efficiently is an opportunity to "play" with numbers and patterns and reinforcing one's number sense.

Things to know about divisibility rules

(a) **Test for divisibility**: The ability to determine whether numbers are divisible by certain divisors is a useful skill. It is highly recommended to know, at least, the divisibility rules for division by 2, 3, 4, 5, 6, 8, 9 and 10.

(b) **Evaluate remainders**: If a number a is not divisible by another number n, this means that $a \div n$ yields a non-zero remainder. Many divisibility rules provide us with more than a criterion for divisibility. They also provide us with a quick and efficient way to evaluate the remainder.

(c) **Underlying principles**: Divisibility rules are not magical. They are based on underlying mathematical principles. Understanding *why* a divisibility rule works is as important as knowing *how* to apply it. Sadly enough, this understanding is hardly ever imparted to elementary school children. As we have stressed repeatedly, it is undesirable to teach techniques without explaining their underlying principles, especially in this case, where the capability of understanding is within children's grasp.

(d) **Links to earlier topics**: Divisibility by composite numbers relates to an earlier topic—prime number factorization. The study of divisibility rules is an excellent opportunity to reveal the power and beauty of assembling existing knowledge for solving new problems.

25.1 The principles underlying divisibility rules

The principles underlying divisibility rules rely on facts that we are already familiar with.

25.1.1 Remainder arithmetic

In Chapter 22, we learned that remainders satisfy arithmetic rules. For example, we have seen the following **remainder addition** rule (p. 76):

$$\text{If}$$
$$\text{Remainder}(a \div n) = x$$
$$\text{and}$$
$$\text{Remainder}(b \div n) = y$$
$$\text{then}$$
$$\text{Remainder}((a+b) \div n) = \text{Remainder}((x+y) \div n)$$

Using the **modulo** operation (p. 68), this can be rewritten in the more compact form,

$$a \bmod n = x \quad \text{and} \quad b \bmod n = y \quad \text{implies} \quad (a+b) \bmod n = (x+y) \bmod n.$$

For example, if we are given that

$$\text{Remainder}(347 \div 6) = 5,$$

and

$$\text{Remainder}(21 \div 6) = 3,$$

then, since $347 + 21 = 368$ and $5 + 3 = 8$, it follows that

$$\text{Remainder}(368 \div 6) = \text{Remainder}(8 \div 6),$$

and the latter is much easier to evaluate—2.

25.1.2 Subtraction of a multiple does not change the remainder

A particular implication of remainder arithmetic is that the remainder does not change if we subtract from the dividend a multiple of the divisor. In formal notations, let a, b, n be natural numbers. Then,

$$\begin{gathered}\text{If} \\ b \text{ is a multiple of } n \\ \text{then} \\ \text{Remainder}(a \div n) = \text{Remainder}((a-b) \div n).\end{gathered}$$

This property is best understood through an example. Suppose we wanted to find the remainder of the following expression,

$$347 \div 6.$$

The remainder will not change if we subtract from 347 a multiple of 6. Since $300 = 50 \times 6$ is a multiple of 6, and since $347 - 300 = 47$, it follows that

$$\text{Remainder}(347 \div 6) = \text{Remainder}(47 \div 6).$$

We have thus reduced the problem to another, simpler problem that yields the same answer.

The subtraction of multiples can be used repeatedly. The remainder will not change if we subtract $7 \times 6 = 42$, which is a multiple of 6, from 47,

$$\text{Remainder}(47 \div 6) = \text{Remainder}(5 \div 6).$$

Since in the latter case, the dividend, 5, is smaller than the divisor, 6, the remainder equals the dividend.

Putting all the pieces together, we conclude that

$$\text{Remainder}(347 \div 6) = 5.$$

25.1.3 Prime number factorization

On p. 95, we have seen that a is divisible by n if and only if the prime factors of n are a subset of the prime factors of a.

For example, 362 is divisible by 6 if and only if its prime factors contain the prime factors of 6. Since

$$\mathcal{P}(6) = \{2, 3\}$$
$$\mathcal{P}(362) = \{2, 181\}$$

the prime factors of 362 do not contain the prime factors of 6, hence 362 is not divisible by 6.

25.2 Divisibility rules

25.2.1 Divisibility by 10

The most rudimentary divisibility rule is that testing for divisibility by 10. Every first grader knows the rule: a number is divisible by 10 if and only if its ones digit is zero. Our goal is to elucidate *why* this divisibility rule works.

To have a concrete example at hand, consider the quotient $3256 \div 10$. We proceed as follows:

(a) A multi-digit number is composed of decimal units,

$$3256 = (3 \times 1000) + (2 \times 100) + (5 \times 10) + (6 \times 1).$$

(Please note: the parentheses are not mandatory; we added them for the ease of visual parsing.)

(b) Every decimal unit from 10 onward is a multiple of 10: 100 is 10×10, $1000 = 10 \times 100$, and so on.

(c) (3×1000) is a multiple of 1000 and 1000 is a multiple of 10. From the transitivity of the multiple-of relation (p. 87), (3×1000) is a multiple of 10. We may deduct (3×1000) from the dividend 3256 without affecting the remainder, hence

$$\text{Remainder}(3256 \div 10) = \text{Remainder}(256 \div 10).$$

(d) (2×100) is a multiple of 100, and 100 is a multiple of 10, hence (2×100) is a multiple of 10. We may deduct (2×100) from the dividend 256 without affecting the remainder,

$$\text{Remainder}(256 \div 10) = \text{Remainder}(56 \div 10).$$

(e) (5×10) is a multiple of 10. We may deduct (5×10) from the dividend 56 without affecting the remainder,

$$\text{Remainder}(56 \div 10) = \text{Remainder}(6 \div 10).$$

(f) The remainder of $6 \div 10$ is 6.

Assembling all the parts together, we conclude that the remainder of $3256 \div 10$ is the same as the remainder of $6 \div 10$. In particular, since 6 is not divisible by 10, neither is 3256.

The important observation is that these considerations were independent of the particular number chosen, i.e., we may view 3256 as a **representative example**. Replacing 3256 by any other number would result in the same conclusion: the remainder obtained when dividing a number by 10 is only determined by the ones digit of that number.

Summarizing:

> When dividing a number by 10, the remainder equals its ones digit. In particular, a number is divisible by 10 if and only if its ones digit is zero.

25.2.2 Divisibility by 5

Divisibility by 5 hinges on the same principles as divisibility by 10. Without further ado, let's consider the arithmetic expression,

$$3256 \div 5.$$

Repeating the same steps as above:

(a) A multi-digit number is composed of decimal units,

$$3256 = (3 \times 1000) + (2 \times 100) + (5 \times 10) + (6 \times 1).$$

(b) 10 is multiple of 5. It follows from the transitivity of the multiple-of relation that every decimal unit from 10 onward is a multiple of 5.

(c) (3×1000) is a multiple of 1000 and 1000 is a multiple of 5, hence (3×1000) is a multiple of 5. We may deduct (3×1000) from the dividend 3256 without affecting the remainder, hence

$$\text{Remainder}(3256 \div 5) = \text{Remainder}(256 \div 5).$$

(d) (2×100) is a multiple of 100, and 100 is a multiple of 5, hence (2×100) is a multiple of 5. We may deduct (2×100) from the dividend 256 without affecting the remainder,

$$\text{Remainder}(256 \div 5) = \text{Remainder}(56 \div 5).$$

(e) (5×10) is a multiple of 10, and 10 is a multiple of 5, hence (5×10) is a multiple of 5. We may deduct (5×10) from the dividend 56 without affecting the remainder,

$$\text{Remainder}(56 \div 5) = \text{Remainder}(6 \div 5).$$

(f) The remainder of $6 \div 5$ is 1.

Testing for divisibility by 5 and 10 differs only in the last stage, when remaining with the ones digit of the number being tested.

Summarizing:

> When dividing a number by 5, the remainder is the same as the remainder when dividing its ones digit by 5. In particular, a number is divisible by 5 if and only if its ones digit is either 0 or 5.

25.2.3 Divisibility by 4

The rule for determining divisibility by 4 is based on the fact that 100 is a multiple of 4 (indeed, $100 = 25 \times 4$). By the transitivity of the multiple-of relation, it follows that all decimal units greater than 100 (i.e., 1000, 10000, etc.) are also multiples of 4. In fact, 1 and 10 are the only decimal units not divisible by 4.

To see how to exploit these facts, consider the following quotient,

$$5973 \div 4.$$

We proceed as follows:

(a) A multi-digit number is composed of decimal units,

$$5973 = (5 \times 1000) + (9 \times 100) + (7 \times 10) + (3 \times 1).$$

(b) (5×1000) is a multiple of 1000 and 1000 is a multiple of 4, hence (5×1000) is a multiple of 4. We may subtract (5×1000) from 5973 without affecting the remainder,

$$\text{Remainder}(5973 \div 4) = \text{Remainder}(973 \div 4).$$

(c) (9×100) is a multiple of 100 and 100 is a multiple of 4, hence (9×100) is a multiple of 4. We may subtract (9×100) from 973, without affecting the remainder,

$$\text{Remainder}(973 \div 4) = \text{Remainder}(73 \div 4).$$

(d) The tens digit cannot be eliminated because as pointed out above, 10 is not a multiple of 4. Yet, finding out whether a two-digit number is divisible by 4 is a much easier task than finding out whether a four-digit number is divisible by 4. In the present case, since 72 is a multiple of 4 we deduce that the remainder of $73 \div 4$ is 1.

(e) Putting it all together,

$$\text{Remainder}(5973 \div 4) = 1.$$

Once again, the procedure is independent of the particular number at hand, hence we can formulate the following divisibility rule:

> When dividing a number by 4, the remainder is the same as the remainder when dividing by 4 the two-digit number formed by its two right-most digits. In particular, a number is divisible by 4 if and only if the two-digit number formed by its two right-most digits is divisible by 4.

If determining whether a given number is divisible by 2, 5 or 10 relies only on the ones digit of that number, determining whether a given number is divisible by 4 relies on its two right-most digits. Take for example the daunting number $10,972,516$. It suffices to inspect its two right-most digits that form the two-digit number 16, to determine that $10,972,516$ is divisible by 4 because 16 is divisible by 4.

25.2.4 Divisibility by 8

Divisibility by 8 is based on similar principles as divisibility by 4: it is based on the fact that 1000 is a multiple of 8 (indeed, $1000 = 125 \times 8$). By the transitivity of the multiple-of relation, all the decimal units greater than 1000 (i.e., 10000, 100000, etc.) are also multiples of 8.

To see how to exploit these facts consider the following quotient,

$$25973 \div 4.$$

We proceed as follows:

(a) A multi-digit number is composed of decimal units,

$$25973 = (2 \times 10000) + (5 \times 1000) + (9 \times 100) + (7 \times 10) + (3 \times 1).$$

(b) (2×10000) is a multiple of 10000 and 10000 is a multiple of 8, hence (2×10000) is a multiple of 8. We may subtract (2×10000) from 25973 without affecting the remainder,

$$\text{Remainder}(25973 \div 8) = \text{Remainder}(5973 \div 8).$$

(c) (5×1000) is a multiple of 1000 and 1000 is a multiple of 8, hence (5×1000) is a multiple of 8. We may subtract (5×1000) from 5973, without affecting the remainder,

$$\text{Remainder}(5973 \div 8) = \text{Remainder}(973 \div 8).$$

(d) The hundreds digit cannot be eliminated, because 100 is not a multiple of 8. Yet, finding out whether a three-digit number is divisible by 8 is a much easier task than finding out whether a five-digit number is divisible by 8. In the present case, we may further observe that 800 is a multiple of 8, and subtract 800 from 973 without affecting the remainder,

$$\text{Remainder}(973 \div 8) = \text{Remainder}(173 \div 8).$$

Moreover, since $160 = 20 \times 8$ is a multiple of 8,

$$\text{Remainder}(173 \div 8) = \text{Remainder}(13 \div 8)$$

and the latter equals 5.

(e) Putting everything together,

$$\text{Remainder}(25973 \div 8) = 5.$$

Once again, the procedure is independent of the particular numbers at hand, hence we can formulate the following divisibility rule:

When dividing a number by 8, the remainder is the same as the remainder when dividing by 8 the three-digit number formed by the three right-most digits. In particular, a number is divisible by 8 if and only if the three-digit number formed by its three right-most digits is divisible by 8.

Once again, take the number 10,972,516. It suffices to inspect its three right-most digits that form the three-digit number 516, to determine that $10{,}972{,}516 \div 8$ results in a remainder of 4, because

$$516 \div 8 = 64 + \text{R}\,(4 \div 8).$$

25.2.5 Divisibility by 9

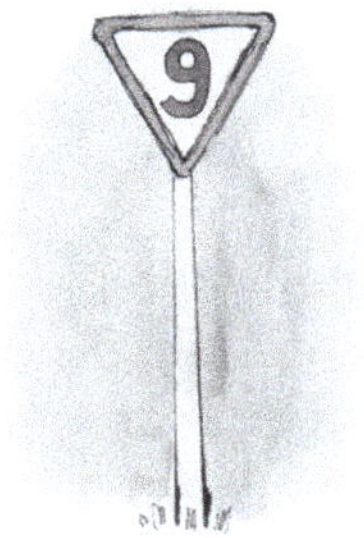

The divisibility rules for division by 2, 4, 5, 8 and 10 have a common property. Divisibility and remainders can be determined by ignoring all but a fixed number of digits (one digit for divisibility by 2, 5 and 10, two digits for divisibility by 4 and three digits for divisibility by 8). As we will see shortly, to test whether a given number is divisible by 9 requires taking into account all of its digits.

The divisibility rule for 9 is based on the fact that the numbers 9, 99, 999, 9999 etc, are all multiples of 9 (we will call these numbers "recurring-9 numbers"). The common property of recurring-9 numbers (except for comprising only the digit 9) is that they are one less than a decimal unit.

This fact can be exploited as follows: Consider for example, the number 34726, which we decompose into decimal units:

$$34726 = (3 \times 10000) + (4 \times 1000) + (7 \times 100) + (2 \times 10) + (6 \times 1).$$

Take each decimal unit, and write is as the sum of a recurring-9 number and one. Applying the distributive property we obtain,

$$3 \times 10000 = 3 \times (9999 + 1) = 3 \times 9999 + 3$$
$$4 \times 1000 = 4 \times (999 + 1) = 4 \times 999 + 4$$
$$7 \times 100 = 7 \times (99 + 1) = 7 \times 99 + 7$$
$$2 \times 10 = 2 \times (9 + 1) = 2 \times 9 + 2$$
$$6 \times 1 = 6.$$

Adding everything up, reordering the various terms,

$$34726 = (3 \times 9999 + 4 \times 999 + 7 \times 99 + 2 \times 9) + (3 + 4 + 7 + 2 + 6).$$

The terms colored in orange are all multiples of 9. Therefore, if we deduct them from 34726, we obtain a number whose remainder upon division by 9 is the same as that of 34726,

$$\text{Remainder}(34726 \div 9) = \text{Remainder}((3 + 4 + 7 + 2 + 6) \div 9).$$

Examine the dividend on the right hand side: it is the sum of the digits of the dividend on the left hand side. Hence, we can formulate the following divisibility rule:

> When dividing a number by 9, the remainder is the same as when dividing by 9 the sum of its digits. In particular, a number is divisible by 9 if and only if the sum of its digits is divisible by 9.

Consider, for example, the eight-digit number 72, 968, 405. Since its digits add up to 41, we deduce that

$$\text{Remainder}(72,968,405 \div 9) = \text{Remainder}(41 \div 9).$$

What is the remainder of $41 \div 9$? We can use the same criterion again, with $4 + 1 = 5$, to obtain that

$$\text{Remainder}(41 \div 9) = \text{Remainder}(5 \div 9),$$

and the latter equals 5.

25.2.6 Divisibility by 3

Since 9 is a multiple of 3, the transitivity of the multiple-of relation implies that all multiples of 9 are also multiples of 3. In particular, recurring-9 numbers are multiples of 3. As a result, divisibility by 3 can be tested in the exact same way as divisibility by 9:

> When dividing a number by 3, the remainder is the same as when dividing by 3 the sum of its digits. In particular, a number is divisible by 3 if and only if the sum of its digits is divisible by 3.

25.2.7 Divisibility by 7

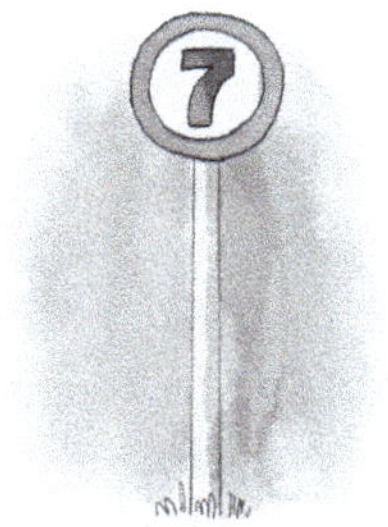

The divisibility rules considered so far are relatively easy to implement because they require considering either a fixed number of digits (divisibility by 2, 4, 5, 8, and 10) or the sum of the digits (division by 3 and 9). Testing for divisibility by 7 is a more complicated task, and as a result is seldom taught in schools. In fact, there is no easy method to test for divisibility by 7, and one finds several methods in the literature, each with its advantages and disadvantages. In this section, two different methods will be presented. Both will be demonstrated in evaluating the remainder of

$$5816 \div 7.$$

25.2.7.1 Method 1: Decomposing $10 = 7 + 3$

Every multi-digit number can be represented as a sum or its ones digit and a multiple of 10. In the present case,

$$5816 = 581 \times 10 + 6.$$

If we decompose the 10 into $7 + 3$ and apply the distributive property, we obtain

$$5816 = 581 \times (7 + 3) + 6 = 581 \times 7 + 581 \times 3 + 6.$$

We do not need to evaluate the term in orange to know that it is a multiple of 7 (every product of 7 is a multiple of 7). Hence, we can deduct it, without changing the remainder. Evaluating the remaining terms, $581 \times 3 + 6 = 1749$, we deduce that

$$\text{Remainder}(5816 \div 7) = \text{Remainder}(1749 \div 7).$$

What did we gain? A smaller dividend. We may now repeat the same procedure: writing

$$1749 = 174 \times 10 + 9 = 174 \times 7 + 174 \times 3 + 9,$$

we deduct the term in orange, evaluate the remaining terms, $174 \times 3 + 9 = 531$, and obtain

$$\text{Remainder}(1749 \div 7) = \text{Remainder}(531 \div 7).$$

Proceeding in the same way,

$$531 = 53 \times 10 + 1 = 53 \times 7 + 53 \times 3 + 1.$$

Deducting the term in orange and evaluating the remaining terms, $53 \times 3 + 1 = 160$, we get

$$\text{Remainder}(531 \div 7) = \text{Remainder}(160 \div 7).$$

One more step,

$$160 = 16 \times 10 + 0 = 16 \times 7 + 16 \times 3 + 0,$$

and since $16 \times 3 = 48$,

$$\text{Remainder}(160 \div 7) = \text{Remainder}(48 \div 7).$$

At this stage we may perform one more step, or exploit our proficiency in the multiplication table to determine that

$$48 \div 7 = 6 + \text{R}(6 \div 7),$$

hence

$$\text{Remainder}(5816 \div 7) = 6.$$

Please note: This divisibility rule is more complex than those considered earlier in two respects:

(1) It requires more steps.
(2) It requires multiplying multi-digit numbers by 3.

25.2.7.2 Method 2: Decomposing all decimal units

The second approach for testing divisibility by 7 is based on the following identities:

$$1 = (0 \times 7) + 1 = 0 + 1$$
$$10 = (1 \times 7) + 3 = 7 + 3$$
$$100 = (14 \times 7) + 4 = 98 + 2$$
$$1000 = (143 \times 7) - 1 = 1001 - 1$$
$$10000 = (1429 \times 7) - 3 = 10003 - 3$$
$$100000 = (14286 \times 7) - 2 = 100002 - 2,$$

where the orange colored terms on the right are multiples of 7.

Consider once again the remainder of the expression

$$5816 \div 7.$$

We start by decomposing 5816 into decimal units,

$$5816 = (5 \times 1000) + (8 \times 100) + (1 \times 10) + (6 \times 1).$$

We apply the distributive property for each decimal unit,

$$5 \times 1000 = 5 \times (1001 - 1) = 5 \times 1001 - 5 \times 1$$

$$8 \times 100 = 8 \times (98 + 2) = 8 \times 98 + 8 \times 2$$

$$1 \times 10 = 1 \times (7 + 3) = 1 \times 7 + 1 \times 3$$

$$6 \times 1 = 6 \times (0 + 1) = 6 \times 0 + 6 \times 1.$$

The orange colored terms on the right hand sides are multiples of 7, hence we may deduct them without affecting the remainder. The remaining terms add up to

$$-(5 \times 1) + (8 \times 2) + (1 \times 3) + (6 \times 1) = 20,$$

hence,

$$\text{Remainder}(5816 \div 7) = \text{Remainder}(20 \div 7),$$

and the latter equals 6.

How did we obtain 20 from 5816? We added together its ones digits, 6, three times its tens digit, 1, and twice its hundreds digits, 8; we then subtracted its thousands digit, 5. In other words, we replaced the number 5816 by a **weighted sum** of its digits, the weights being (+1) for the ones digit, (+3) for the tens digit, (+2) for the hundreds digit and (−1) for the thousands digit. If there were more digits, the weights of the next digits would be (−3) for the ten-thousands digits, (−2) for the hundreds-thousands digits, and then back to (+1) for the millions digit, (+3) for the ten-millions digit, and so on.

In summary:

> When dividing a number by 7, the remainder is the same as when dividing by 7 the weighted sum of its digits, where the weights, starting from the ones digits are $(1, 3, 2, -1, -2, -3, 1, 3, 2, \ldots)$. In particular, a number is divisible by 7 if and only if the weighted sum of its digits is divisible by 7.

25.2.8 Divisibility by 11

Divisibility by 11 is based on the following identities:

$$1 = (0 \times 11) + 1 = 0 + 1$$
$$10 = (1 \times 11) - 1 = 11 - 1$$
$$100 = (9 \times 11) + 1 = 99 + 1$$
$$1000 = (91 \times 11) - 1 = 1001 - 1$$
$$10000 = (909 \times 11) + 1 = 9999 + 1$$
$$100000 = (9091 \times 11) - 1 = 100001 - 1,$$

where the orange colored terms on the right are multiples of 11.

Consider for example the remainder of the expression

$$54812 \div 11.$$

We start by decomposing the dividend into decimal units,

$$54812 = (5 \times 10000) + (4 \times 1000) + (8 \times 100) + (1 \times 10) + (2 \times 1).$$

We apply the distributive property for each decimal unit,

$$5 \times 10000 = 5 \times (9999 + 1) = 5 \times 9999 + 5$$
$$4 \times 1000 = 4 \times (1001 - 1) = 4 \times 1001 - 4$$
$$8 \times 100 = 8 \times (99 + 1) = 8 \times 99 + 8$$
$$1 \times 10 = 1 \times (11 - 1) = 1 \times 11 - 1$$
$$2 \times 1 = 2 \times (0 + 1) = 2 \times 0 + 2.$$

The orange colored terms on the right hand sides are multiples of 11, hence we may deduct them without affecting the remainder. The remaining terms add up to

$$5 - 4 + 8 - 1 + 2 = 10,$$

hence,

$$\text{Remainder}(54812 \div 11) = \text{Remainder}(10 \div 11),$$

and the latter equals 10.

How did we obtain 10 from 54812? By taking the **alternating sum** of its digits: we added together all its digits at the odd positions (ones, hundreds, ten-thousands) and subtracted from that sum its digits at the even positions (tens and thousands). This example leads us to the following generalization:

> When dividing a number by 11, the remainder is the same as when dividing by 11 the alternating sum of its digits. In particular, a number is divisible by 11 if and only if that alternating sum is divisible by 11.

25.3 Divisibility by composite numbers

At this stage, we have learned divisibility rules for division by 2, 3, 4, 5, 7, 8, 9, 10 and 11. Since every natural number is divisible by 1, there is no need for a divisibility rule for division by 1. Thus, we have learned to test divisibility by all number up to 11, except for 6.

This section studies divisibility rules for division by composite numbers. At the end of this section, we will have the tools necessary to determine if a number is divisible, for example, by 12, 20 or 45. All those divisibility rules are based on the one principle mentioned earlier: a number a is divisible by n if and only if the prime factors of n are a subset of the prime factors of a.

25.3.1 Divisibility by 6

The prime factors of 6 are,

$$\mathcal{P}(6) = \{2, 3\},$$

therefore, a number is divisible by 6 if and only if 2 and 3 are among its prime factors. Stated differently:

> A number is divisible by 6 if and only if it is divisible by both 2 and 3.

Since we already know how to test for divisibility by both 2 and 3, we have all the necessary tools to determine whether a number is divisible by 6. For example, to determine whether 5412 is divisible by 6, we need to check whether it is divisible by 2 (it is, because its ones digit is even) and whether it is divisible by 3 (it is, because the sum of its digits, 12, is divisible by 3).

Please note: This divisibility rule only determines if a number is divisible by 6, but does not reveal the remainder if that number is not divisible by 6. In this context, see Problem 25.5 below.

25.3.2 Divisibility by 4

As shown above, we can determine whether a number is divisible by 6 by performing two separate tests: by checking whether is it divisible by 2 and by 3. This is so because 2 and 3 are the prime factors of 6. A natural question is whether this always works: can we always determine whether is number is divisible by a composite number by performing separate divisibility tests, one for each prime factor of the divisor.

The answer turns out to be negative. To understand what can go wrong, let's revisit divisibility by 4. The prime factors of 4 are

$$\mathcal{P}(4) = \{2, 2\}.$$

That is, 4 has a prime factor that recurs twice. If we tried to naively mimic the way we determine divisibility by 6, we would say that *"a number is divisible by* 4 *if and only if it is divisible by* 2 *and divisible by* 2*"*. This is absurd! What's the point of checking twice that a number is divisible by 2? To determine whether a number is divisible by 4 we need to check whether the prime factor 2 recurs at least twice in its prime number factorization. Checking twice whether that number is even does not achieve that goal.

We conclude that the method exemplified in the divisibility by 6 rule only works as long as the prime factors composing the divisor are all distinct. When it comes to divisibility by 4, we have to resort to the method described on p. 128.

25.3.3 Divisibility by 12

Consider next divisibility by 12. The prime factors of 12 are

$$\mathcal{P}(12) = \{2, 2, 3\}.$$

Thus, a number is divisible by 12 if and only if its prime factors contain the prime factor 2 at least twice and the prime factor 3 at least once.

Even though the prime factors of 12 are not all distinct, we can determine whether a number is divisible by 12 in two separate stages:

(1) Determine whether its prime factors contain the prime factor 2 at least twice.
(2) Determine whether its prime factors contain the prime factor 3 at least once.

The first stage amounts to determining whether that number is divisible by 4, whereas the second stage amounts to determining whether that number is divisible by 3.

Summarizing:

> A number is divisible by 12 if and only if it is divisible by both 4 and 3.

25.4 Validation using divisibility tests

Until several decades ago, calculators did not exist and the evaluation of the four arithmetic operations was an important skill in daily life. Because long calculations are prone to errors, it was important to develop efficient methods to locate errors. A popular validation method was based on divisibility rules.

Consider, for example, the following arithmetic expression,

$$13,297 \times 5,477.$$

In Chapter 19, we learned how to evaluate such products using the standard algorithm. This is a long and tedious evaluation, that even an experienced adult could easily perform erroneously.

Suppose we tried to evaluate this product and obtained the following result:

$$66,324,569.$$

How can this result be verified?

Most people would try to repeat the calculation and see whether the same answer is obtained. The drawback of this approach is that if an error is due to a misconception rather than, say, a typographical error, the same error is likely to recur.

We now present a validation method based on divisibility tests. First, recall the following remainder arithmetic rule (p. 78):

$$\begin{gathered}\text{If}\\ \text{Remainder}(a \div n) = x\\ \text{and}\\ \text{Remainder}(b \div n) = y\\ \text{then}\\ \text{Remainder}((a \times b) \div n) = \text{Remainder}((x \times y) \div n).\end{gathered}$$

If we replace a by 13,297, b by 5,477 and n by 9 we get

$$\text{If}$$
$$\text{Remainder}(13{,}297 \div 9) = x$$
$$\text{and}$$
$$\text{Remainder}(5{,}477 \div 9) = y$$
$$\text{then}$$
$$\text{Remainder}((13{,}297 \times 5{,}477) \div 9) = \text{Remainder}(x \times y) \div 9).$$

Once we recall that we have a simple and efficient way to evaluate the remainder for division by 9, we realize that that last sentence is far less daunting that it may seem first. Since

$$\text{Remainder}(13{,}297 \div 9) = 4,$$

and

$$\text{Remainder}(5{,}477 \div 9) = 5,$$

we deduce that

$$\text{Remainder}((13{,}297 \times 5{,}477) \div 9) = \text{Remainder}(5 \times 4) \div 9),$$

and the latter is easily evaluated—2.

With very little effort we determined a property of the sought product: if we divide it by 9 we get a remainder of 2. We can then verify whether our candidate solution satisfies this requirement. The digits of 66,324,569 sum up to 41, hence

$$\text{Remainder}(66{,}324{,}569 \div 9) = \text{Remainder}(41 \div 9),$$

and the later is 5—not 2. Thus, we know at once that our answer was erroneous.

Please note:

(1) Had we found that the candidate solution had the correct remainder upon division by 9, this would not have guaranteed the correctness of the calculation. After all, every ninth number yields the same remainder when divided by 9. This method of verification *may* find errors, but is not guaranteed to find them.
(2) Why use a divisibility-by-9 criterion? The largest the divisor, the more likely we are to discover errors. 9 is the largest number having a divisibility rule that is easy to implement.

Mathematical problems

Problem 25.1. Check whether 23,076,991 is divisible by 2, 3, 4, 5, 6, 7, 8, 9, 10, 11, and 12. In each case, find the remainder if it is not divisible.

Problem 25.2. Determine whether each one of the following claims is correct:

(a) $5 \times 5 \times 3 \times 3$ is divisible by 11,
(b) 21×21 is divisible by 9,
(c) 717×55 is divisible by 25,
(d) $17 \times 17 \times 17$ is the square of a whole number,
(e) $9 \times 9 \times 9$ is the square of a whole number.

Problem 25.3. Let a be an even number not divisible by 4. What are the possible remainders if we divide it by 8?

Problem 25.4. Let a be an odd number not divisible by 3. What are the possible remainders if we divide it by 6?

Problem 25.5. Determine the remainder obtained when dividing a number by 6 in each of the following cases:

(a) That number is even and divisible by 3.
(b) That number is even and when divided by 3 yields a remainder of 1.
(c) That number is even and when divided by 3 yields a remainder of 2.
(d) That number is odd and divisible by 3.
(e) That number is odd and when divided by 3 yields a remainder of 1.
(f) That number is odd and when divided by 3 yields a remainder of 2.

Problem 25.6. Evaluate the following product,

$$325 \times 501$$

using the standard multiplication algorithm. Validate your solution using the validation method described on p. 139.

Problem 25.7. Propose a divisibility rule for division by 45 based on divisibility by smaller numbers.

Problem 25.8. Propose a divisibility rule for division by 24 based on divisibility by smaller numbers. Does it predict that 96 is divisible by 24? Does it predict that 108 is divisible by 24?

Chapter 26

Sequences

A sequence is an **ordered set** of items—a set of items that can be matched in a one-to-one manner to the natural numbers. This is a formal way of saying that one of the items is assigned the number 1 and designated as **first**, another is assigned the number 2 and designated as **second**, and so on. A sequence may comprise a finite set of items, in which case one of the items is designated as **last**, or, it may comprise an infinite number of items. In that case, to every natural number n corresponds one and only one item located at the n-th place.

When we say that a sequence is ordered, we mean that the items in the sequence "inherit" the order of the natural numbers. For example, we say that the item that is the seventh on the list comes after the item that is the fourth on the list, because 7 is greater than 4. Likewise, an item is the **successor** of an other item if the natural number assigned to it is the successor of the number assigned to that other item.

Generally, a sequence may comprise all sorts of items. For example, a list of children ordered lexicographically is a sequence. In elementary school mathematics, though, one mostly encounters **number sequences** and sequences of geometric objects.

Opportunities:

(a) Playing with number sequences is a golden opportunity to practice arithmetic.
(b) Pattern detection is at the heart of mathematical endeavor.
(c) Number sequences are abundant in natural phenomena.
(d) Sequences display interesting connections between number patterns and geometry.

26.1 Some examples

Consider the following list of numbers,

$$2, 5, 8, 11, 14, 17.$$

The order of precedence, as usual from left to right, imposes an order among those numbers—this is a number sequence, comprising 6 elements. You surely have noticed that this sequence follows a simple rule: each element, starting from the second, is greater than its predecessor by 3.

The following list of numbers,

$$4, 17, 325, 12209, 0, 102,$$

is also a finite number sequence. Unlike the previous sequence, there is no apparent rule describing the order of those numbers.

Consider next the following sequence

$$2, 4, 6, 8, 10, \dots$$

Here, five numbers are displayed, followed by an ellipsis. The ellipsis indicates that the sequence continues indefinitely—this is an infinite sequence. How can we know which element comes at the sixth place? The ellipsis notation is only used in cases where the sequence has a clear regularity, so that it is self-evident how to generate the next elements of the sequence. In the present case, it is self-evident that this sequence comprises all the even numbers in increasing order, which means that each element is greater than its predecessor by 2. Thus, the next elements on the list are $12, 14, 16, \ldots$ (and here we use ellipses again).

Why are sequences interesting?

Sequences are usually not mentioned explicitly in school curricula, yet, they are worth studying all along the K-6 years for numerous reasons:

(a) Mathematics is above all the art of structure, patterns and rules. Studying sequences is an excellent opportunity for practicing the identification of patterns.

(b) Skip counting, which is the generative rule of a certain type of number sequences, connects addition to multiplication.

(c) Mathematics is full of fascinating examples of both number sequences and sequences of geometric objects. The study of such sequences opens up a wealth of enrichment topics that are built upon curricular content and reveal many mathematical gems.

26.2 Arithmetic progressions

Consider the following infinite sequences:

$$3, 5, 7, 9, 11, \ldots$$

$$14, 20, 26, 32, 38, \ldots$$

$$100, 97, 94, 91, 88, \ldots.$$

These three sequences share a common property: the difference between every two succeeding elements is always fixed. In the first sequence, each element is greater than its predecessor by 2, that is, the difference between every two succeeding elements is 2. In the second sequence, each element is greater than its predecessor by 6. In the last sequence, each element is smaller than its predecessor by 3. A sequence in which there is a fixed difference between every two succeeding elements is called an **arithmetic progression**.

One may wonder what is "arithmetic" about an arithmetic progression. Arithmetic progressions are called so since each element, starting from the second, is the **average** of its predecessor and its successor. This average, or **mean**, is formally called an **arithmetic mean**, to distinguish it from other types of averages (see below). It is the arithmetic mean that gave its name to the arithmetic progression.

26.2.1 Finding the element at a given location

Consider the arithmetic progression,

$$21, 25, 29, 33, 37, \ldots,$$

whose first element is 21 and whose fixed difference is 4. What is the sixth element in this sequence? What is the seventh element? What is the hundredth element?

Since we know the rule generating this sequence, we can use a systematic method to find any desired element. Each element is greater than its predecessor by 4, hence the element following 37, which is the sixth in the sequence, is $37 + 4 = 41$. The seventh element is then $41 + 4 = 45$, and so on. In principle, we can proceed indefinitely, finding along the way the hundredth element in the sequence.

This way of proceeding is of course inefficient, tiresome, and totally impractical if we want to find, say, the millionth element in the sequence. This is where mathematics calls for creative thinking.

To figure out a systematic way for finding the element at any desired location, consider the following characterization of the first few items in the sequence:

$$\begin{aligned}
1\text{st element} &= 21 &&= 21 + (0 \times 4)\\
2\text{nd element} &= 21 + 4 &&= 21 + (1 \times 4)\\
3\text{rd element} &= 21 + 4 + 4 &&= 21 + (2 \times 4)\\
4\text{th element} &= 21 + 4 + 4 + 4 &&= 21 + (3 \times 4)\\
5\text{th element} &= 21 + 4 + 4 + 4 + 4 &&= 21 + (4 \times 4).
\end{aligned}$$

In the same manner, the 6th element is obtained by adding 5×4 to 21, the 7th element is obtained by adding 6×4 to 21, and so on. Having elucidated the rule, you can easily evaluate now the hundredth element and the millionth element:

$$100\text{th element} = 21 + (99 \times 4)$$

$$1{,}000{,}000\text{th element} = 21 + (999{,}999 \times 4).$$

At this stage, you have certainly realized that there is nothing special about the numbers 21 and 4. The same way of proceeding can be applied regardless of the particular values of the first element and the difference between succeeding elements. In fact, we can generalize the rule into a formula: In an arithmetic progression,

$$n\text{-th element} = 1\text{st element} + (n-1) \times \text{difference between succeeding elements.}$$

Such a formula is typically learned in high school. I believe that with some guidance, elementary school students might actually be able to *discover* it.

26.2.2 Arithmetic series

There is a very famous story about the German mathematician Carl Friedrich Gauss. According the legend, when Gauss was 7, his teacher gave the class an assignment—to evaluate the sum of all the natural numbers between 1 and 100. In other words, he required them to evaluate the following sum:

$$1 + 2 + 3 + 4 + \cdots + 99 + 100.$$

The teacher hoped, so the legend says, to keep his pupils busy while he could take a nap. However, young Gauss gave this a moment's thought before voicing the answer, 5050, leaving the teacher speechless.

How did 7-year-old Gauss add up 100 numbers that fast? Instead of adding the natural numbers between 1 and 100 one after the other, he first reordered the addends. He noticed that the first and last elements add up to 101. Likewise, the second element and the element that is second to the last add up to 101, and so on. Thus, he summed up the natural numbers between 1 and 100 in the following order:

$$(1 + 100) + (2 + 99) + (3 + 98) + \cdots + (50 + 51).$$

The natural numbers between 1 and 100 can be paired into 50 pairs all having the same sum—101. Therefore, Gauss concluded that this sum equals

$$50 \times 101 = 5050.$$

Kaboom! Whether this legend is based on a true story on not, it shows us once more the power of mathematics: creative thinking can save a great deal of work.

This method for evaluating sums, attributed to Gauss, is not limited to sums of succeeding natural numbers. The natural numbers between 1 and 100 are just an instance of an arithmetic progression starting at 1, in which the fixed difference between succeeding elements is 1. More generally, we may evaluate the sum of the elements of any arithmetic progression. Such a sum is called an **arithmetic series**.

Consider the following finite arithmetic progression,

$$3, 7, 11, 15, 19, 23, 27, 31, 35,$$

consisting of 9 elements starting at 3, and a difference of 4 between succeeding elements. What is the sum of those 9 elements?

Repeating Gauss' method, we proceed by adding together the first and last elements, the second and second to last elements, and so on, leading to the following scheme:

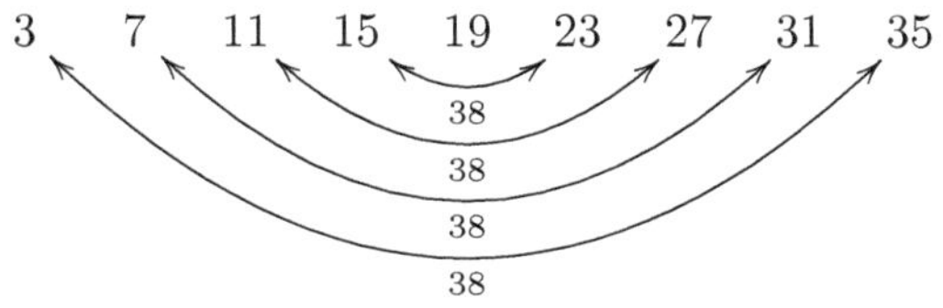

Thus, every pairs of numbers that are matched together yield the same sum, 38. Since there are 9 numbers in the sequence, we could form 4 such pairs, leaving one element, 19, unmatched to another element. This arithmetic series is therefore equal to

$$4 \times 38 + 19 = 171.$$

Let us turn your attention to two interesting points. First, why does adding pairs of elements equally distanced from the sequence end points always yield the same sum? This is a characteristic of arithmetic progressions. Since every element is greater than its predecessor by the same amount as it is less than its successor, the sum of the second and the second to last elements must be the same as the sum of the first and the last element. Second, notice that the element that remained unmatched, 19, is exactly half the fixed sum, 38, of the pairs of elements that were matched together. I urge you to figure out why this will always be the case for a sequence comprising an odd number of elements.

The method displayed above for evaluating an arithmetic series comprising 9 elements is somewhat disappointing, even though it is efficient. The fact that we had to account for an element unmatched to another element spoiled to some extent the beauty of Gauss' legendary solution. A more elegant solution strategy should be insensitive to whether the sequence comprises an even or odd number of elements.

An alternative approach, even more ingenious, for calculating arithmetic series is the following: rather than adding up the elements of the sequence once, *we add them up twice!* "What can be so enlightening in more work?" you may ask. The idea is to arrange the two progressions in opposite order, as follows:

3	7	11	15	19	23	27	31	35
35	31	27	23	19	15	11	7	3

Then, we add up the first element of the sequence with the first element of the reversed sequence, the second element of the sequence with the second element of the reversed sequence, and so on:

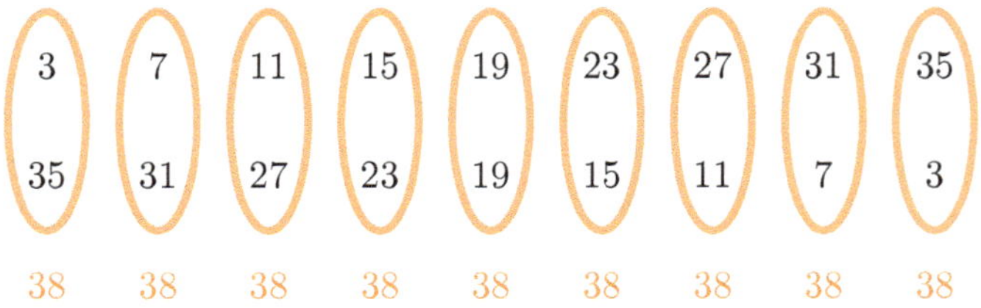

The combined sums of the two sequences is then the number of elements in the original sequence, 9, multiplied by the sum of the first and last elements of the original sequence, 38, i.e.,

$$9 \times 38 = 342.$$

But this sum is twice larger than the series we want to evaluate—a series comprising the element of just one sequence. Hence, we need to divide the above product by 2, yielding once again the desired sum,

$$342 \div 2 = 171.$$

Note that in this alternative approach, we do not care wether the sequence comprises an odd or even number of elements.

This calls for formulating a general rule. In any arithmetic series,

$$\text{Series} = \#\ \text{elements} \times (\text{first element} + \text{last element}) \div 2.$$

This method for evaluating an arithmetic series can also be endowed with a geometric interpretation. If we interpret a number n as the area of a rectangle whose dimensions are n-by-1 length units, then the series we evaluated equals the area in square units of the following shape,

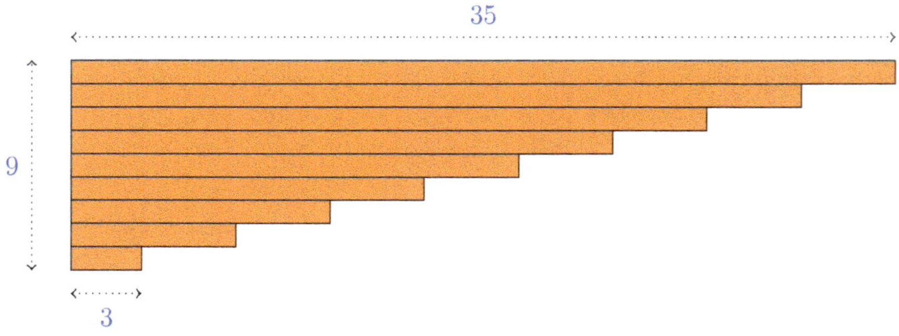

where the shortest rectangle has dimensions 3-by-1 and the longest rectangle has dimensions 35-by-1.

Rather than adding up numbers to calculate the area of this shape, we replicate it, and combine the two copies into a rectangle, as follows:

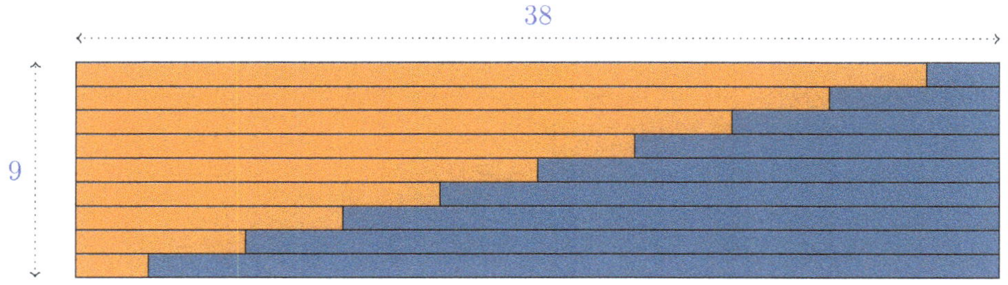

Since the area of this rectangle is 9×38 square units, the area of the original shape is half of it. Please note the similarity of this approach to the way we find the area of a right triangle (Volume 1, p. 233).

26.3 Geometric progressions

Consider the following two infinite sequences:

$$3, 6, 12, 24, 48, \ldots$$
$$2, 8, 32, 128, 512, \ldots.$$

These two sequences share a common property: the ratio between every two succeeding elements is fixed. In the first sequence, every element equals twice its predecessors, that is, the ratio between every two succeeding elements is 2. In the second sequence, every element equals 4 times its predecessor. A sequence in which there is a fixed ratio between every two succeeding elements is called a **geometric progression**.

Geometric progressions are called so because each element, starting from the second, is the **geometric mean** of its predecessor and its successor. The geometric mean of two numbers is the square root of their product. For example, the geometric mean of 3 and 12 is

$$\sqrt{3 \times 12} = \sqrt{36} = 6.$$

You may wonder what is "geometric" about a geometric mean. A possible explanation is that the geometric mean of two numbers is the side length of a square whose area is the same as the area of the rectangle whose dimensions are the two given numbers. Indeed, a square whose side length is 6-units has the same area as a 3-by-12 unit rectangle.

26.3.1 Exponential growth

Geometric progressions are considered harder to work with compared to arithmetic progression, because multiplication is harder to evaluate than addition. Yet, many children may find interest in the dramatic rate of increase of geometric progressions such as

$$2, 4, 8, 16, 32, \ldots.$$

Geometric progressions in which the fixed ratio between succeeding elements is greater than one are said to **grow exponentially**.

26.3.2 Exponential decay

The sequence

$$\frac{1}{2}, \frac{1}{4}, \frac{1}{8}, \frac{1}{16}, \ldots$$

is also geometric progression. Here, each element is half of its predecessor, so that the fixed ratio between succeeding elements is a fraction–1/2.

Unlike the previous examples, here each element is smaller than its predecessor. Geometric progressions in which the fixed ratio between succeeding elements is less than one are said to **decay exponentially**. Exponential decay is "fast"—the 20-th element in this sequence is already less than a millionth. Exponentially decaying geometric progressions can be taught in school alongside with fraction multiplication.

26.3.3 Geometric progressions in daily life

Geometric progressions are widespread in our lives. Below are two occurrences of exponentially growing sequences:

(a) Suppose you deposit \$1000 in your bank account for an annual interest rate of 10%. The balance after one year will be \$1100, obtained by multiplying the starting sum by 1.1. After another year, an additional 10% will accrue to the 1100; that is, the balance after two years will be equal to the first year's balance times 1.1. At the end of each year, the balance is the product of the previous year's balance and 1.1. The sequence of balances at the end of each year forms a geometric progression.

(b) Bacteria reproduce by division, at the end of which a single bacterium turns into two bacteria. Imagine a type of bacterium that divides once an hour. Beginning with a single bacterium, there will be 2 bacteria after an hour, 4 bacteria after two hours, and so on. At this pace, after 24 hours there will be more than 16,000,000 bacteria.

26.4 Square numbers

Consider the following sequence,

$$1, 4, 9, 16, 25, 36, \ldots$$

This sequence comprises the squares of the natural numbers, i.e., the n-element in the sequence is $n \times n$. Children are acquainted with this sequence, as it fills up the main diagonal of the multiplication table. In fact, I would recommend knowing by heart all square numbers up to 20×20,

$$\ldots, 49, 64, 81, 100, 121, 144, 169, 196, 225, 256, 289, 324, 361, 400.$$

The sequence of square numbers has interesting properties. For example, consider the difference between any two succeeding elements:

Surprisingly, the difference between succeeding numbers forms an arithmetic progression of succeeding odd numbers. An explanation to this intriguing discovery is depicted below:

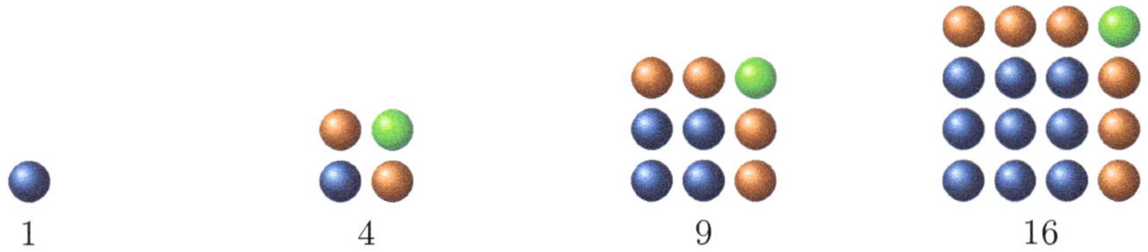

In the above figure, every square number $n \times n$ is represented by a square array of n-by-n dots. To obtain the next number in the sequence, i.e., $(n+1) \times (n+1)$, we form a square array of $(n+1)$-by-$(n+1)$ dots by appending two rows of n dots along two of the square's sides (in orange), plus one more dot at the corner (in green). This diagram reveals in a very illustrative way a fact learned in high school algebra: the difference between $(n+1) \times (n+1)$ and $n \times n$ is $2n+1$.

Thus, the sequence of differences between two succeeding square numbers is a sequence whose n-th element is $2n+1$, i.e., a sequence of odd numbers starting at 3.

26.5 Triangular numbers

Next, consider the following sequence:

$$1, 3, 6, 10, 15, 21, 28, 36, 45, 55, \ldots$$

The rule governing this sequence may not be easily discerned. Note that the differences between succeeding elements generates the arithmetic progression,

$$2, 3, 4, 5, 6, 7 \ldots.$$

This sequence of numbers also bears a geometric interpretation. Consider the following sequence of patterns,

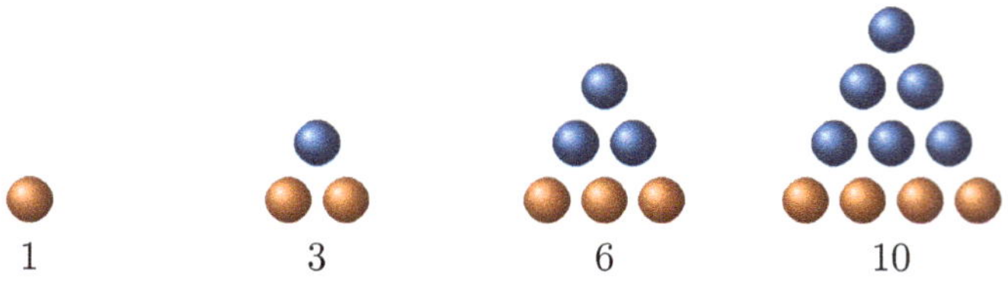

The number sequence can be given a concrete representation as a sequence of dots arranged in a triangular pattern; thus the name **triangular numbers**. The n-th number in the sequence is represented by a triangular arrangement that has n dots at the bottom and is n dots high. We colored the dots in such a way that makes evident that the n-th number is greater than the $(n - 1)$-st number by n.

Another property of the sequence of triangular numbers is that it we add together every two succeeding numbers, we obtain the following sequence,

$$4, 9, 16, 25, 36, \ldots,$$

that is, square numbers. Every square number is the sum of two succeeding triangular numbers. I urge you to investigate by geometric considerations how this miracle occurs.

As advocated at the beginning of this chapter, the discovery of patterns and rules in sequences is an opportunity for developing curiosity and creativity, and a great source of *aha!* moments. Connections between arithmetic and geometry are of particular value in the development of intuition and the search for out of the box solutions.

26.6 The Fibonacci sequence

Consider the following sequence,

$$1, 2, 3, 5, 8, 13, 21, 34, 55, \ldots$$

Someone seeing this sequence for the first time may have a hard time discerning its rule. The property of this sequence is that each element, starting from the third, is the sum of its two predecessors. For example, the fifth element, 8, is the sum of the third element, 3 and the fourth element, 5.

This sequence first appeared in ancient Indian writings, but it achieved prominence in 1202 in a book written by the Italian mathematician Leonardo of Pisa, also known as Fibonacci.

Fibonacci got interested in the sequence, now bearing his name, while trying to explain the reproductive rate of rabbits. Fibonacci proposed the following model for reproductivity: At the start, assume a population of rabbits consisting of one male and one female, i.e., one couple. Assume that each month this pair of rabbits gives birth to a new pair of rabbits, a male and female. Furthermore, each new pair of rabbits gives birth to a pair of rabbits at the rate of once a month, but only starting from the second month of its life. Another assumption made by Fibonacci was that rabbits never die, nor become infertile.

Having adopted this simplified model for reproductivity, Fibonacci made the following observations:

(a) At the end of the first month, the population of rabbits will comprise 2 pairs of rabbits—the original pair and an additional one that was just born,
(b) At the end of the second month, the population of rabbits will comprise 3 pairs of rabbits, since only the original pair of rabbits was fertile at the beginning of the month. Yet, by now there are already two pairs of fertile rabbits.
(c) At the end of the third month, the population of rabbits will comprise 5 pairs—the 3 that were present at the end of the previous month, plus 2 produced by the 2 fertile pairs. By now there are 3 pairs of fertile rabbits.
(d) At the end of the fourth month, the population of rabbits will comprise 8 pairs—the 5 that were present at the end of the previous month, plus 3 produced by the 3 fertile pairs. By now there are 5 pairs of fertile rabbits.

Continuing to count monthly, the number of pairs of rabbits forms a Fibonacci sequence.

26.6.1 Fibonacci numbers in daily life

In addition to being used as a coarse model for rabbit reproduction, Fibonacci numbers occur in many natural contexts. For example, the number of petals in flowers tends to be Fibonacci numbers. Lilies have three petals, buttercups have five petals, the chicory has 21 petals and daisies often have 34 or 55 petals.

Next, consider the following sequence of geometric shapes. We start with two unit squares juxtaposed to each other, forming a 2-by-1 rectangle.

Attaching a square to the longer side of the rectangle, we obtain a 2-by-3 rectangle:

Attaching a square to the longer side of the rectangle, we obtain a 5-by-3 rectangle:

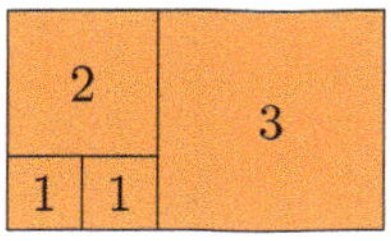

Attaching a square to the longer side of the rectangle, we obtain a 5-by-8 rectangle:

Attaching a square to the longer side of the rectangle, we obtain a 13-by-8 rectangle:

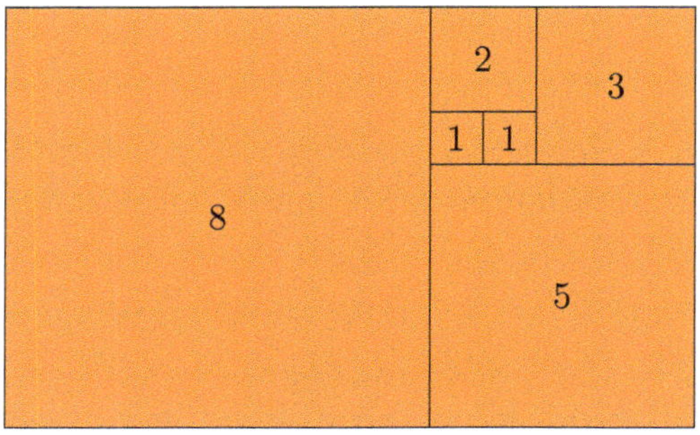

Attaching a square to the longer side of the rectangle, we obtain a 13-by-21 rectangle: This construction gives a sequence of increasing rectangles, whose side lengths are consecutive Fibonacci numbers. If we now form a curve connecting opposing vertices of squares, we obtain a spiral:

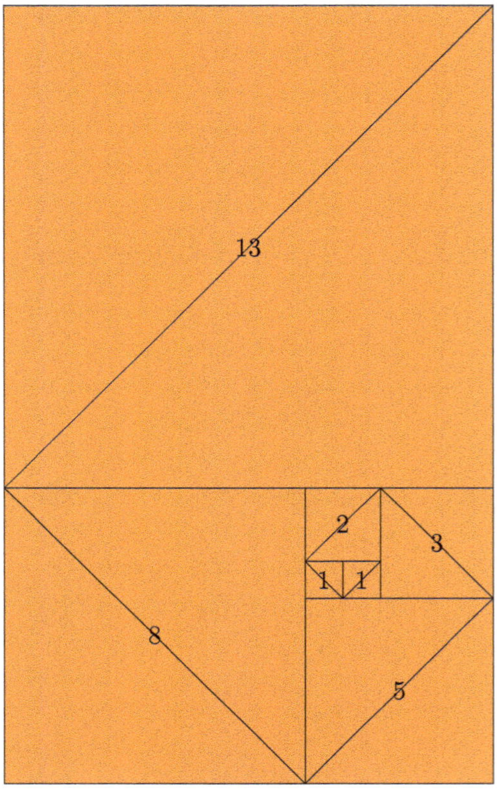

It turns out that many snails and conch shells have spiral structures similar to this spiral, or some version of it.

26.6.2 The golden ratio

The Fibonacci sequence is an increasing sequence whose elements grow very fast. Thus, we might be tempted to speculate that the Fibonacci sequence is in fact a geometric progression.

A simple way to verify whether a sequence is a geometric progression is to calculate the ratio of every two succeeding elements and check whether it is constant. The following table displays the ratio between succeeding elements in the Fibonacci sequence:

n	n-th element	$(n+1)$-st element	ratio
1	1	2	2
2	2	3	1.5
3	3	5	1.66666
4	5	8	1.6
6	8	13	1.625
7	13	21	1.61538
8	21	34	1.61905
9	34	55	1.61764

The Fibonacci numbers, as it turns out, do not form a geometric progression, as the ratio between succeeding elements is not fixed. Yet, it is "almost" a geometric progression. As we proceed along the sequence, the ratio between succeeding elements approaches a limit—a number roughly equal to 1.6180339877. This number is called the **golden ratio**, and like the Fibonacci sequence, has preoccupied scientists, and even artists, for hundreds of years.

26.7 The isoperimetric problem

Consider the following question:

> Of all the rectangles having a 40cm long perimeter, which one has the largest area?

Problems of this type, where one has to find the maximal area while constrained by a fixed perimeter, are known as **isoperimetric problems**. This geometric problem turns out to have interesting arithmetic aspects.

First, think of what you already know about rectangles having a 40cm long perimeter. Since a rectangle has two pairs of equal sides, it follows that the total length of two adjacent sides is half of the perimeter, 20cm. For example, if the length of one side is 6cm, then the length of the other side is 14cm, and if the length of one side is 19.5cm, then the length of the other side is 0.5cm.

Among all those rectangles, there is one that stands out—the square whose sides are 10cm long (remember that a square is a rectangle). The area of that square is 100sq-cm. If we now elongate one of its sides, while maintaining a constant perimeter, then the other side must be shortened by the same amount. The question is whether the area can be enlarged in this manner.

The table below presents areas of various rectangles having a perimeter of 40cm. In principle, side lengths should not be limited to whole numbers, but we have our own reasons to consider only rectangles of integer dimensions at this time.

Side length (cm)	Other side length (cm)	Area (sq-cm)
10	10	100
11	9	99
12	8	96
13	7	91
14	6	84
15	5	75

This table suggests that among all rectangles having a 40cm long perimeter, the square is the one having the largest area. Increasing the **aspect ratio** of the rectangle (the proportional relationship between its length and its width) results in area reduction.

This table reveals, however, much more than just a solution to the isoperimetric problem posed at the beginning of this section. Consider the extent by which the area decreases as we increase the aspect ratio of the rectangle:

$$100 - 99 = 1$$
$$100 - 96 = 4$$
$$100 - 91 = 9$$
$$100 - 84 = 16$$
$$100 - 75 = 25$$

The sequence of area reductions is a sequence of square numbers. How did that happen? Also, is it something rooted in our choice a 40cm long perimeter?

There is one simple way to refute that last proposition—consider another example. Consider rectangles having, say, a 48cm long perimeter. If all its sides are equal, we obtain a square whose sides are 12cm long, and whose area is 144 sq-cm. Changing its dimensions, one cm at a time, we obtain the following table:

Side length (cm)	Other side length (cm)	Area (sq-cm)
12	12	144
13	11	143
14	10	140
15	9	135
16	8	128
17	7	119

The extents by which the areas decrease as we increase the aspect ratio of the rectangle are,

$$144 - 143 = 1$$
$$144 - 140 = 4$$
$$144 - 135 = 9$$
$$144 - 128 = 16$$
$$144 - 119 = 25$$

Again, the sequence of square numbers.

This surprising phenomenon can be elucidated using algebraic tools, but the challenge is to explain it using more rudimentary tools, accessible to elementary school children. Let's try to answer the following question: suppose we start with a square—we already know that it does not matter what its dimensions are—and transform it into a rectangle by elongating one side by 3cm and reducing the other side by 3cm. Why should the area decrease by 9 sq-cm?

The answer to that riddle is depicted below:

On the left is a square of arbitrary dimension. In order to elongate one of its sides by 3cm, we cut out a 3cm-wide strip, rotate it, and place it adjacent to the other side of the rectangle, thus obtaining the rectangle on the right. In doing so, there is a piece that has to be left out. This is the 3cm by 3cm square colored light blue. As a result, the area has decreased by 9 sq-cm. This illustration explains why the reduction in area always corresponds to the area of a square, whose side length coincides with the extent by which the sides of the square were elongated or reduced.

26.8 The sequence of factorials

Consider the following sequence,

$$1, 2, 6, 24, 120, 720, \ldots$$

Can you identify the rule?

The second element is equal to twice the first element, the third element is equal to 3 times the second element, the fourth element is equal to 4 times the third element, and so on. Equivalently, we may rewrite this sequence as follows:

$$1, \quad 1 \times 2, \quad 1 \times 2 \times 3, \quad 1 \times 2 \times 3 \times 4, \ldots.$$

The n-th element in this sequence is the product of all natural numbers between 1 and n. This product is called the **factorial** of n, and it denoted by $n!$. For example.

$$7! = 1 \times 2 \times 3 \times 4 \times 5 \times 6 \times 7 = 5040.$$

The sequence of factorials grows very fast. For example, the next four elements after 720 are

$$5,040, \quad 40,320, \quad 362,880, \quad 3,628,800.$$

Factorials play a very important role in combinatorics (Volume 1, p. 88). We will demonstrate it through an example accessible to elementary school children.

Suppose you have two different color blocks: red and green, and you want to build a tower using exactly one block of each color. How many different towers could you build?

The answer of course is 2. Either the red block is at the bottom and the green block is on top of it, or the other way around, the green block is at the bottom and the red block is on top of it:

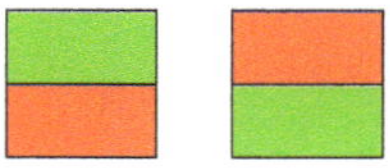

Assume that a blue block is added. How many different towers can be built with three different color blocks? After a short trial and error session, you will surely reach the conclusion that there are 6 different towers as depicted below:

The transition from 2-block towers to 3-block towers can be made systematic as follows: Take the two different 2-block towers. To each one of them add a blue block—once at the bottom, once in the middle, and once on top. Thus, each of the two 2-block towers can be made into three different 3-block towers, which is why we end up with 3×2 different 3-block towers.

So let's add a purple block. How many different towers can be built with four different color blocks? This time, blind trial and error is likely to lead you astray. There are too many combinations to keep track of. Instead, we repeat the argument that lead us from

2-block to 3-block towers. We already know that there are six different 3-block towers. Take each of them, and add to it a purple block. The purple block can be placed in four different locations, hence each of the 3-block towers produces four different 4-block towers. The total number of 4-block towers is therefore, $4 \times 6 = 24$. Proceeding this way, we conclude that n different blocks give rise to $n!$ different towers.

Activity: Sitting on a bench

Rather than working with blocks, the following activity can be performed in class:

(a) Place a bench in front of the class, and invite 2 children to the front of the class. In how many different ways the 2 children can be seated. This question can be solved easily.

(b) Invite a third child and repeat the question. It might take the class some time to exhaust all possible arrangements until reaching the conclusion that there are 6 different possibilities.

(c) Invite a fourth child and repeat the question. At this stage, one must find ways to systematically keep track of arrangement that were already listed. Let the children devise their own strategies. You can also lead them toward the solution, for example, by suggesting "lets first count all the arrangements where Joe is seated on the left".

(d) To verify that the children understood the solution strategy, ask them to find out how many seating arrangements there are for 5 children.

(e) If the class is attentive, you may end up discovering that 10 children can be seated in more than 3 million different arrangements.

Mathematical problems

Problem 26.1. Consider the following sequence,

$$81, 78, 75, 72, 69, \ldots.$$

(a) What is the pattern in this sequence?

(b) What is the 6th element in this sequence? What is the 19th element?

(c) What is the sum of the first 9 elements?

Problem 26.2. Find an efficient method to evaluate the sum of all the numbers appearing in the multiplication table.

Chapter 27

Fractions

Fractions are considered *the* major stumbling block in children's elementary mathematical education. The first encounter with fractions (and arithmetic operations involving fractions) is often a turning point, after which many children start viewing mathematics as an alien discipline, remote from their daily life routine. In order to survive in the classroom, children end up adopting a how-to attitude, memorizing mantras such as *just invert and multiply*. The resulting fraction trauma lingers well into high school and beyond: many adults regard fractions as an obscure subject. Ask around: how many of your friends are able to attribute any meaning to, say, TWO-FIFTHS divided by THREE-QUARTERS?

Just like in every other subject, the key to mastering fractions is having a sound foundation. The goal of this chapter—the first in a series of six chapters devoted to fractions—is to clarify what fractions are, and in what sense they extend our number system, which comprised first only natural numbers.

Hurdles:

(a) Children have difficulties internalizing that fractions are numbers just as natural numbers.
(b) Children have difficulties understanding the symbolic representation of fractions, and the disparate meanings of the denominator and the numerator.
(c) An incorrect conceptualization of fractions it at the core of difficulties with arithmetic operations with fractions.

Remedies:

(a) Use fractions in daily contexts as early as possible.

(b) Introduce fractions in a broad variety of contexts, both as parts of a whole and parts of a quantity; be creative in varying the wholes, or units to which fractions refer. Recall that an abstract concept is best internalized through a continual exposure to varied concrete realizations.

(c) Expose children to the numeral representation of fractions as early as possible; don't be afraid to create confusion as you weren't afraid to write multi-digit numbers before teaching the principles of the decimal representation.

(d) Use number names that best reflect the meaning of fractions; e.g., prefer TWO-THIRDS upon TWO-OVER-THREE, which only alludes to the fraction's numeral representation.

Opportunities:

The early stages of learning fractions are crucial for building up intuition; a good conceptualization of the fraction concept will be rewarding for years to come.

27.1 Natural numbers and fractions

Mankind invented the natural numbers to measure, or quantify sets of objects (see Chapter 1). In ancient prehistory, natural numbers were always attached to the nouns they were quantifying. In a sense, natural numbers served as machines replicating nouns: *five trees* was just shorthand for *tree, tree, tree, tree, and tree.* It took a long way, and a great deal of abstraction, until mankind formed a number concept detached from the noun it quantifies. Along with the abstraction of numbers, came the invention of arithmetic operations, which are abstractions of concrete operations, such as joining together, taking apart, etc. A central goal of early mathematical education is helping children acquire the conceptualization of the natural numbers within the span of several years, an evolutionary process that once took tens of thousands of years.

If natural numbers are used to quantify **pluralities of a whole**, fractions are used to quantify **parts of a whole**. The need to quantify part of something arises naturally in daily contexts. Just as with natural numbers, it took humankind generations to invent an abstract measure of a part, detached from the whole it is part of. The conceptualization of a fraction is the core of solid foundations; it is a process requiring time, guidance and a continual exposure to a plethora of concrete examples.

27.1.1 Natural numbers as quantifiers

Numbers—natural numbers and fractions alike—are used for **quantifying** concrete objects. Exactly as we started our mathematical journey by using numbers for quantifying sets, i.e., for counting, so the first skill one should acquire when it comes to fractions is using them for quantifying parts of a whole.

One of my guiding principles in teaching fractions is to invoke, as often as possible, the traits they share with the natural numbers. Thus, our first step will be to remind ourselves how natural numbers are used as quantifiers. To this end, consider the following dialogue between two characters, Hannah and Tami, who will accompany us throughout this chapter:

;

HANNAH: Tami, how many apples are there?

TAMI: I see an apple, an apple, an apple and an apple.
HANNAH: So how many apples are there?
TAMI: ONE, ONE, ONE, and ONE. In mathematical notation:

$$1 + 1 + 1 + 1.$$

HANNAH: Your answer is correct but cumbersome, and it would have been impossible to make sense of it if there were many more apples.

It is precisely in order to prevent such awkwardness that the word FOUR was invented. FOUR is synonym for ONE, ONE, ONE and ONE. Numbers provide us with a means to answer the question "*how many?*" in a single word.

There is a key matter which I want to draw your attention to. Since FOUR is just short-hand for ONE, ONE, ONE and ONE, the meaning of FOUR derives from the meaning of ONE. There can be no FOUR without a ONE!

When we use natural numbers and perform arithmetic operations with natural numbers, it is always assumed, either explicitly or implicitly, that there is an underlying notion of a *unit*, or a *whole* or a ONE. The transition from abstract numbers to concrete quantities occurs when one specifies what this ONE is. In the above dialog, for example, the ONE is a single apple; in other contexts, the ONE could refer to a crate full of apples, jars of apple sauce, major cities on the East Coast, computer companies or something else having nothing to do with apples.

27.1.2 Fractions as quantifiers

Let's return to our two protagonists:

HANNAH: Tami, how many apples are there?

TAMI: Not even one apple.
HANNAH: Then, *what part* of an apple is there?
TAMI: I see a part of an apple obtained when cutting up one apple into 4 equal parts and keeping just one of those parts.
HANNAH: Then, how much is it?

[*At this stage, Tami remembers that a quotient expresses the size of a share when dividing a dividend into equally sized parts (she must have read Chapter 8). Here, the dividend is* 1 *apple and the divisor is* 4. *Thus, she replies:*]

TAMI: There is $1 \div 4$ of an apple.
HANNAH: There you go again, answering in arithmetic expressions. I want to know how much of an apple there is. I don't want to do arithmetic.

The common trait of the two dialogs, is that Tami's answers were arithmetic expressions. Her first answer, $1 + 1 + 1 + 1$, expresses the outcome of *joining together* a whole, a whole, a whole and a whole. Her second answer, $1 \div 4$, expresses the size of a share obtained by *dividing* a whole into four equal parts. In both cases, the whole happened to be an apple.

In both cases, however, Hannah expected a one-word answer. Precisely as the number FOUR is shorthand for ONE, ONE, ONE and ONE, so the number ONE-QUARTER is shorthand for *the size of a share when dividing a whole into* 4 *equal parts*. A number such as ONE-QUARTER is called a **fraction**.

On fractions and their numeral notation

(a) Numbers such as ONE-QUARTER quantify a part of a whole—they are **fractions**. The word *fraction* originates from the Latin word *fractionem*, which means *a breaking*. Fractions are called so because they represent quantities that can be viewed as breaking a whole into parts.

(b) Fractions, like natural numbers, are quantifiers—they assign a measure, or a size to collections of objects, or to parts of objects. In either case, whether we use natural numbers or fractions, the notion of a *whole*, or a ONE must be defined. Natural numbers indicate *how many* wholes there are, and fractions indicate *how much* of a whole there is. Please note that while the English language distinguishes between *how many?* and *how much?*, depending whether the noun is countable or non-countable, there are languages in which the same word is used for both.

(c) Every natural number is represented by a **numeral**—a string of digits. Fractions, ONE-QUARTER for example, are also represented by numerals. Interestingly, the way we denote fractions alludes to the operation defining them. If ONE-QUARTER is the size of a share when dividing 1 whole into 4 equal parts, then we denote ONE-QUARTER as ONE divided by FOUR. Rather than using the division sign, as in $1 \div 4$, we write this expression vertically,

$$\frac{1}{4}.$$

The dividend is placed above the divisor and the two are separated by a line segment, called a **fraction bar**. In this context, the dividend is called a **numerator** and the divisor is called a **denominator**. The meaning of these terms will be clarified below. For typographical reasons we will sometime replace the vertical notation of fractions by a horizontal notation, slanting the fraction bar, as in $^1/_4$.

(d) It follows that the identity

$$1 \div 4 = \frac{1}{4}$$

is not the outcome of a calculation, but rather a definition. The number on the right-hand side is *defined* as the value of the arithmetic expression on the left-hand side. Please note that this is not the only occurrence of numerals alluding to the operation defining them. The same occurs with negative numbers. The numeral (-3), for example, represents the outcome of subtraction: (-3) is defined as the value resulting from *zero minus three.*

(e) Our notation for fractions combines pairs of natural numbers—a numerator and a denominator. This fact contributes to the widespread misconception that fractions, unlike natural numbers, are not "plain" numbers. Our notation for fractions is indeed more involved than our notation for natural numbers;

moreover, as will be shown in the next chapter, different fractions can have the same value. Yet, fractions are numbers just as natural numbers. Moreover, in the next chapter we will see that natural numbers have equivalent representations as fractions.

(f) Alternative notations for representing parts of a whole exist, for example, **decimals** (see Volume 3). A decimal representation, unlike the fractional representation, does not allude to the division operation that breaks the whole into parts.

(g) In a number system comprising only natural numbers, the result of an operation such as $1 \div 4$ is expressed by a quotient and remainder (see Chapter 22)

$$1 \div 4 = 0 + \mathrm{R}(1 \div 4).$$

Read this equation aloud: when you divide a whole into four equal parts, the size of each share is zero, and there is a remainder of one whole that ought to be divided into four equal parts. In simple words, the division operation was *not* carried out. The incorporation of fractions into our number system provides us with the means to preform every integer division (more on that in the next chapter).

(h) The incorporation of fractions into a number system comprising originally only natural numbers is called in mathematical jargon an **extension** of the number system. The set of numbers comprising both natural numbers and fractions is called the set of (positive) **rational numbers**. This is the first in a series of number system extensions: More extensions of the number system are to follow in higher grades, with the introduction of **negative numbers**, **irrational numbers** and **complex numbers**.

27.2 Unit fractions

27.2.1 Definition

Similarly to ONE-QUARTER quantifying the size of a share when a whole is divided into 4 equal parts, we define fractions expressing the size of a share when dividing a whole into any number of equal parts. Those fractions are called **unit fractions**, because they express a quantity obtained by dividing a unit into equal parts.

The following table presents the first elements of the sequence of unit fractions:

Name	Meaning	Notation
ONE-HALF	Quotient of 1 divided by 2	$^1/_2$
ONE-THIRD	Quotient of 1 divided by 3	$^1/_3$
ONE-QUARTER	Quotient of 1 divided by 4	$^1/_4$
ONE-FIFTH	Quotient of 1 divided by 5	$^1/_5$
ONE-SIXTH	Quotient of 1 divided by 6	$^1/_6$
...	...	...

The numerator of a unit fraction is always 1. The denominator corresponds the number of parts into which the whole is partitioned.

A notational remark: throughout this book, numbers are usually written in symbols, rather than in words (we prefer to write 26 rather than TWENTY-SIX). We will amend this practice when it comes to fractions. Whenever a fraction appears within the text (in contrast to being within a displayed equation), we will use number names, e.g., TWO-THIRDS and TEN-TWELFTHS rather than their symbolic representations, $^2/_3$ and $^{10}/_{12}$.

There is one major reason for preferring number names when it comes to fractions: symbols such as $^2/_3$ can be read out in several ways: TWO-THIRDS, TWO-OVER-THREE or TWO-DIVIDED-BY-THREE. The choice of a number name often triggers a certain perception of its meaning (e.g., TWO-THIRDS implies 2 times ONE-THIRD, whereas TWO-OVER-THREE only reflects a typographic arrangement of two numbers). In what follows, I will define fractions in a certain way, and the use of fraction names ensures that fractions are read out in a way reflecting this definition.

27.2.2 Quantifying with unit fractions

The first fundamental mathematical skill developed at an early age is counting—using natural numbers to quantify the size of sets. It is only through repeated and versatile practice in determining "how many items are there?" that children eventually form in their mind's eye the abstract notion of a *number*. On both historical and pedagogical grounds, the properties of natural numbers and the properties of arithmetic operations on natural numbers all derive from the nature of counting. As stressed repeatedly in Volume 1, if you can count, then you can add, subtract, multiply and divide.

Analogously, when it comes to fractions, the first mathematical skill that needs developing is determining how much of a whole there is, or, incorporating the word *fraction* into the question, what fraction of a whole there is. As with natural numbers, the properties of fractions and the properties of arithmetic operations on fractional operands derive from the ability to quantify using fractions.

The ability to use fractions as quantifiers does not come out of thin air; it requires the same sort of repeated and versatile practice as the ability to use natural numbers as quantifiers. While the use of natural numbers starts at a young age, long before reaching elementary school, this is not so for fractions. Thus, under the pressure of having to cover a full year's curriculum, this crucial stage in one's acquaintance with fractions is often too short, resulting in weak foundations.

The rest of this section presents examples requiring the use of unit fractions to quantify "what fraction of a whole is there?":

Example 1: An apple was divided equally among 3 children. What fraction of an apple did every child get?

Solution: Each child got a share obtained by dividing an apple into 3 equal parts. By definition, each child got ONE-THIRD of an apple. Note that the whole in this case is an apple.

Example 2: What fraction of the disc is blue?

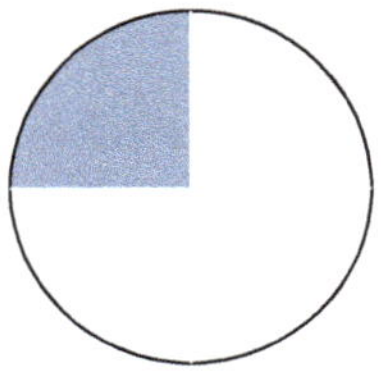

Solution: The blue region (called a **sector**) is a share obtained by partitioning the disc into 4 equal sectors. Therefore, it constitutes ONE-QUARTER of the disc.

Example 3: What fraction of the following figure is black?

;

Solution: Part of the figure is white and part of it is black. Since, by symmetry, the black part and the white part are of equal size, we conclude that the black part constitutes ONE-HALF of this figure.

Example 4: Captain Hook shared a treasure with 4 pirates. Each pirate received an equal share of the treasure, but Captain Hook took twice as much for himself. What fraction of the treasure did each pirate receive? How would your answer change if you knew that the treasure contained 1000 golden coins?

Solution: In order to distribute the treasure as requested, Captain Hook needs to first divide the treasure into 6 equal parts, giving out one part to each pirate (thus, every pirate gets the same share), and keeping two parts for himself (thus, getting twice as much as each of his accomplices): It follows that each pirate got ONE-SIXTH of the treasure.

The whole to which the question alludes is a treasure—not a golden coin. The answer ONE-SIXTH remains valid regardless of the content of that treasure.

Example 5: A family ordered 2 pizzas, partly covered with toppings, as shown below:

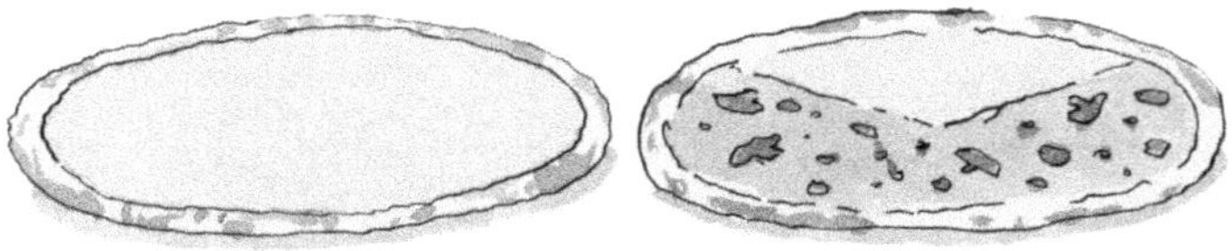

What fraction of the pizzas is covered with toppings?

Solution: Here one needs to be alert and ask: what is the whole whose fraction we are seeking? The answer lies in the question: the whole is the entire order—2 pizzas—not a single pizza.

Take then the whole—the 2 pizzas—and cut out the part covered with toppings. The remaining part is twice as large as the part covered with toppings. This means that the part covered with toppings is a share obtained when dividing the whole into 3 equal parts. Hence, it is ONE-THIRD of the whole.

Example 6: What fraction of this group of children are wearing hats?

Solution: Here again, tread with caution, and start by identifying the whole. The question asks about a *fraction of a group*. We are not asked how many children wear hats; the whole is not a child; it is a group of children. Since the hat-wearing children happen to form a group obtained by dividing the group into 3 equal sub-groups, we conclude that they form ONE-THIRD of that group.

We emphasize the fact that this answer can be given without reference to products of fractions and natural numbers. Later on, one might attribute to this word problem the number sentence,

$$\frac{1}{3} \times 6 = 2.$$

Yet, writing this number sentence is unnecessary for the current purpose: the quantification of parts of a whole.

Example 7: What fraction is displayed in the following figure?

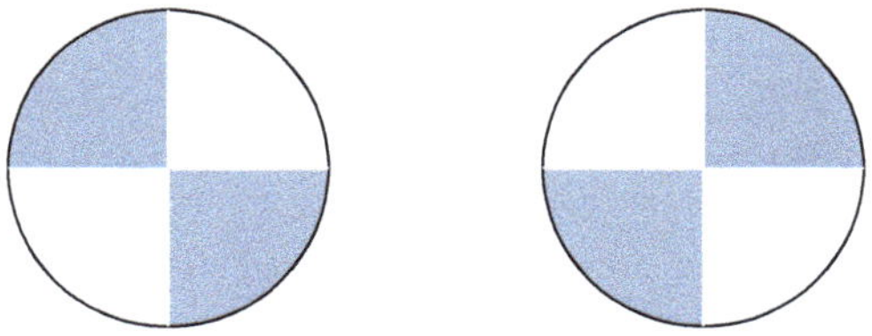

Solution: It is impossible to answer this question since it is not specified which part of what whole one should consider. You could detect in this figure 2 circles, 8 sectors, 1 whole circle colored blue, ONE-HALF of the figure colored blue, or ONE-HALF of the figure colored white.

27.3 General fractions

27.3.1 Definition

As this stage, we have a number system including natural numbers and unit fractions, which we can use to answer *how many* wholes there are, or *how much* of a whole there is. Yet, unit fractions do not suffice for quantifying *every* part of a whole. The number system needs to be extended beyond unit fractions, as demonstrated in the following dialog:

HANNAH: What fraction of the following disc is blue?

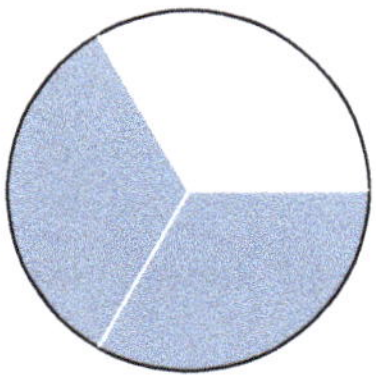

[*Tami realizes at a glance that the blue portion is less than a whole disc, however, it is not a share obtained by dividing the disc into equal parts. In fact, she perceives that the blue portion is composed of 2 parts, each being* ONE-THIRD *of the disc.*]

TAMI: The blue part is 2 times ONE-THIRD of the disc.

HANNAH: If so, what fraction of the disc is blue?

TAMI: ONE-THIRD and ONE-THIRD. Since I learned to denote unit fractions, I can express my answer in mathematical notation:

$$\frac{1}{3} + \frac{1}{3} \qquad \text{or} \qquad 2 \times \frac{1}{3}.$$

HANNAH: There you go again.

Tami will not be able to satisfy Hannah's expectations, because she still lacks the necessary vocabulary. The blue part of the disc is neither an integer amount of discs, nor a unit fraction part of a disc. Tami needs a name for the part of the whole she rightly identifies as ONE-THIRD plus ONE-THIRD. Such a name exists: it is simply TWO-THIRDS. Its numeral representation is,

$$\frac{2}{3}.$$

Fractions that are not unit fractions are multiples, or repeated additions of a unit fraction. The **denominator**, the number under the fraction bar, designates the unit fraction that is being repeatedly added. The **numerator**, the number above the fraction bar, designates the number of unit fractions composing the general fraction.

Let m and n be natural numbers. The fraction $^m/_n$ represents a part of a whole obtained by adding together m copies of $^1/_n$ of that whole. Since $^1/_n$ of a whole is a share obtained when partitioning that whole into n equals parts, we may say that the fraction $^m/_n$ represents part of a whole obtained by dividing the whole into n equals parts and taking m parts out of the n parts.

The definition of the fraction $^m/_n$ can be expressed as a number sentence,

$$\frac{m}{n} = \underbrace{\frac{1}{n} + \frac{1}{n} + \cdots + \frac{1}{n}}_{m \text{ times}},$$

or

$$\frac{m}{n} = m \times \frac{1}{n}.$$

For example,

$$\frac{4}{7} = \frac{1}{7} + \frac{1}{7} + \frac{1}{7} + \frac{1}{7} = 4 \times \frac{1}{7}.$$

Such identities are usually taught in fifth and sixth grades under the scopes of fraction addition and fraction multiplication. We stress again that these identities are introduced as *definitions*; nothing here is being calculated.

27.3.2 Quantifying with general fractions

As with unit fractions, the first skill that needs to be acquired with general fractions is quantification—identifying parts of a whole. Below are some examples.

Example 1: What fraction of the disc is blue? What fraction is orange?

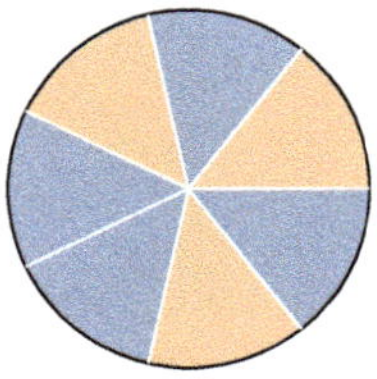

Solution: The disc is partitioned into 7 sectors of equals size, hence each sector is ONE-SEVENTH of the disc. Since there are 4 blue sectors, they constitute FOUR-SEVENTHS of the disc. Since there are 3 orange sectors, they constitute THREE-SEVENTHS of the disc.

Example 2: Anat divided an apple into 5 equal parts. She gave one to Gili, another to Alon and another to Tal. The rest she kept for herself. What fraction of an apple did Gili get? What fraction of an apple did Anat get? What fraction of an apple did Gili and Anat get together?

Solution: Since Anat divided an apple into 5 equal parts, each part is ONE-FIFTH of an apple. This is also the fraction of an apple that Gili got. Since there were originally 5 parts, of which Anat gave out 3, she kept 2 parts for herself, which constitute TWO-FIFTHS of an apple. Finally, together Gili and Anat got 3 parts, which constitute THREE-FIFTHS of an apple.

The last question relates to **fraction addition**. However, we can solve it by using only the definition of a fraction.

Example 3: What fraction of the birds are not red?

Solution: There are 18 birds, out of which 12 are not red. In the next chapter, we will learn that 12 out 18 are TWO-THIRDS, but at this stage we want to stick to methods that rely exclusively on identifying fractions from their definition.

We observe, by mental experimentation, that we could partition this flock of birds into three sets—each constituting ONE-THIRD of the flock—one of which comprising red birds, and the two other comprising birds that are not red:

We conclude that the red birds constitute TWO-THIRDS of the flock.

27.3.3 The rational numbers

On p. 168, we mentioned the term **rational numbers** as designating all those numbers that are either natural numbers or fractions (to be precise, we limit our attention here to positive rational numbers). The term *rational* derives from the Latin word *ratio*, which traces back to the ancient Greek word *logos*, which means one quantity divided by another. Rational numbers had a prominent role in ancient Greek mathematics; see p .177.

There are infinitely many fractions

We have already discussed the infinitude of the natural numbers (Volume 1, p. 9). What about fractions? Are there infinitely many of them?

The answer is yes. In fact, we can find a one-to-one matching between the natural numbers and the unit fractions, assigning the unit fraction $1/n$ to every natural number n. Using mathematical jargon, the set of natural numbers and and the set of unit fractions are both infinite, but have the same **cardinality** (Volume 1, p. 12).

In addition to the unit fractions, there are also infinitely many fractions that are not unit fractions. In fact, we can find a one-to-one matching between the natural numbers and fractions whose numerator is 2, we can find a one-to-one matching between the natural numbers and fractions whose numerator is 3, and so on. Thus, there seems to be an infinitude of infinitely many fractions. Surprisingly, it turns out that it is possible to find a one-to-one matching between the natural numbers and *all* rational numbers, meaning that both sets have the same cardinality. Contrary to simple intuition, there are not "more" fractions than there are natural numbers. (This fact is by no means elementary; it is only in the second half of the 19th century that **set theory** matured to allow for the comparison of infinite sets.)

27.4 Fractions in early history

In Chapter 1, we described the prehistoric development of natural numbers as quantifiers of sets comprising **discrete** objects. Soon enough, there was a need to quantify **continuously**-varying amounts, such as the milk-yield of a cow, the length of a fence or the area of a field. Take length for example: the first step taken by man was to **discretize** measures of length by adopting a **measuring unit**, such as a yardstick. Placing yardsticks end-to-end, it was possible to count how many times that unit is contained in the measured object. This turned the measurement of a continuous entity into that of counting discrete objects.

Nevertheless, this practice had a drawback. More often than not, this measuring practice would leave a **remainder**, or a **residue**. It is quite probable that in early days, residues were simply ignored (as an evidence, the ratio between the circumference of a circle and its diameter is approximated by 3 in the Biblical *Books of Kings*). Yet, sooner or later, along with the agricultural revolution and the development of trade, there was a need to quantify residues (e.g., residues of gold, when measuring the length of a necklace, were not something one could just ignore).

A natural solution (which nevertheless took millennia to develop) is the following: take the measuring unit, say a yardstick, and cut it up into a sub-unit (e.g., cutting up a yard 3-fold yields a sub-unit called a foot). Then, measure how many times the sub-unit fits into the residue. For example, it might so happen that a sub-unit that is contained in the original unit 3 times, is contained in the residue exactly 2 times. Or perhaps, a sub-unit that is contained in the original unit 13 times, is contained in the residue exactly 8 times. Thus, the problem of measuring a residue was converted into that of finding *two* natural numbers, m and n, such that a sub-unit that is contained in the original unit n times is contained in the residue exactly m times. It was the Hindus who originally denoted the outcome of such a measurement by writing those two natural numbers one on top of the other, $^m/_n$, as in our contemporary notation.

Can any residue be quantified this way? For a long time, the answer was believed to be positive, as long as we allow for arbitrarily small sub-units (i.e., n can be as large as we please). The first proof that "unmeasurable" residues exist is attributed to the Greek mathematician Hippasus of Metapontum (5th century BC), who allegedly showed that the diagonal of a square cannot be measured as an exact fraction of its side. We will return to this pivotal event in the history of mathematics in Volume 3, when we present **irrational numbers**.

27.5 Fractions as part of a quantity

We defined fractions as numbers quantifying **parts of a whole**. Yet, one often encounters the saying that fractions, unlike natural numbers, have multiple meanings, notably as quantifying **part of a quantity**. Thus, you may rightly ask yourself whether there is something I was hiding from you in my introduction to the fraction concept.

The answer is no: I was not hiding anything. Fractions quantify parts of a whole. But as stated on p. 166, the whole could be almost any entity. The only constraint, when referring to a fraction of a whole, is that the whole be something that can be partitioned. *Whole* could refer to a pizza, 10 pizzas, ONE-HALF of a pizza, all the books in a bookstore, and so on. It is in those instances where the whole is a collection of "other wholes" (e.g., the whole is 18 birds as in Example 3 on p. 175), that one refers to fractions as quantifying part of a quantity.

When the whole is a collection of discrete objects, we often want to alternate between the knowledge of *what fraction of the whole* and the knowledge of *how many objects*. Referring again to Example 3 on p. 175, TWO-THIRDS of the flock of birds corresponds to 12 birds. Please note that the numbers TWO-THIRDS and 12 quantify different wholes—a flock of birds in the first case and a single bird in the second case. We will return to this type of correspondences in Chapter 31, but at this point I want to stress once again: a fraction, as an abstract concept, has a unique definition. Its concrete representation depends on the nature of the *whole* to which it refers.

As an epilog, let's meet our two protagonists, Hannah and Tami, for one last time:

HANNAH: Tami, I bought a box of 12 oranges, which I will be happy to share with you.
TAMI: Thanks, Hannah.
HANNAH: How much would you like?
TAMI: Could I have a half?
HANNAH: Sure, take half an orange.

This dialog is quite innocent until the last line, where we realize that Tami and Hannah have a very different perception of what ONE-HALF refers to. As explained above, the concrete meaning of ONE-HALF depends on the concrete meaning of the whole. When Tami asks for ONE-HALF, the whole she has in mind is a box full of oranges. When Hannah hands Tami ONE-HALF of an orange, the whole she has in mind is one orange. There was room for ambiguity because the question "How much would you like?" did not determine in any clear way in what *units* the answer should be given, i.e., what does ONE represent—a box or an orange? Remember this dialog whenever fractions confuse you. There is a good chance that your confusion is due to the fact that the whole is not well-defined.

Mathematical problems

Problem 27.1. Complete the following table. Each line exhibits a different whole.

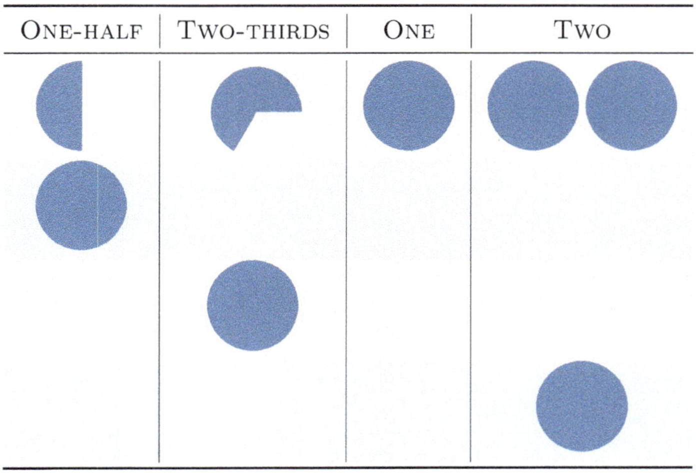

Problem 27.2. Determine, without performing any calculation, which fraction of the large square is blue:

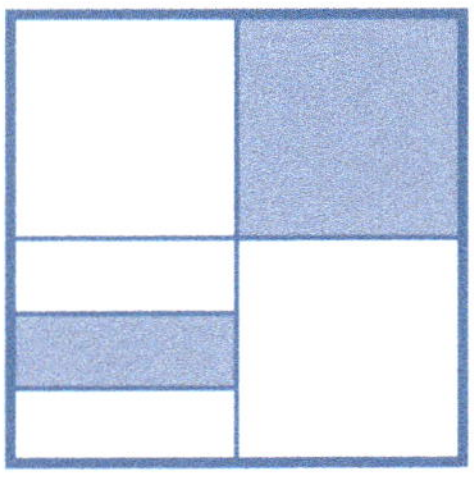

Problem 27.3. Determine, without performing any calculation, which fraction of the large square is blue:

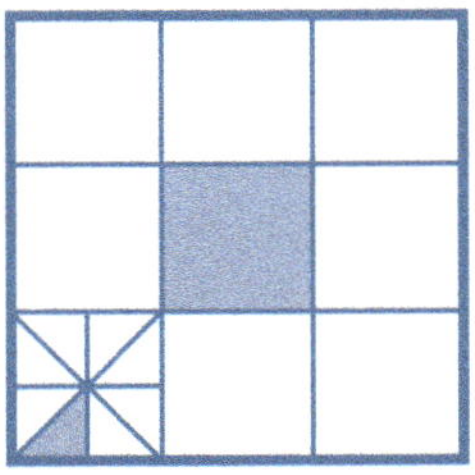

Pedagogical problems

Problem 27.4. Ms. Merz asked Gili to color ONE-HALF of a rectangle. Gili colored part of a rectangle as illustrated below:

Did Gili fulfill the required task? Explain your answer.

Chapter 28

Quotients of Natural Numbers

In Chapter 27, we introduced **fractions**, which together with the natural numbers form the extended set of **rational numbers**. Fractions quantify **parts of a whole**. They are represented by numerals of the form $^{m}/_{n}$, where both the **denominator** n and the **numerator** m are natural numbers. The fraction $^{m}/_{n}$ is defined as m times the size of a share obtained when partitioning a whole into n equal parts.

This chapter expands our acquaintance with fractions, focusing on two main subject matters:

(a) In a number system including fractions, every two natural numbers can be divided, resulting in a quotient that is either a whole number or a fraction. One might suggest that even without fractions, every two natural numbers can be divided, resulting in a quotient plus a **remainder**. However, as described in Chapter 22, a non-zero remainder implies partial, or incomplete division. In an extended number system comprising fractions, the division operation regains its original interpretation: the quotient is the size of a share when *the entire* dividend is divided into parts of equal size.

(b) Two *distinct* fractions, $^{m}/_{n}$ and $^{k}/_{l}$, can have the *same* value, i.e., represent the same number. In such case, we say that the two fractions are **equivalent**. That two different numerals can represent the same value is a new concept, which takes time and practice to internalize. Multiple uses of equivalences between fractions will be presented in the following chapters. These observations lead to the general fact that a number can be represented by an infinite numbers of fractions.

Hurdles:

(a) Children are confused as to the meaning of non-unit fractions; does TWO-THIRDS mean 2 times ONE-THIRD, or is it the quotient of 2 divided by 3?

(b) Children don't understand that two equivalent fractions are different numeral representations of the *same* number.
(c) Children adopt the misconception that fractions are always less than 1.

Remedies:

(a) Be systematic—define non-unit fractions as the repeated addition of unit fractions; demonstrate the relation between non-unit fractions and quotients of natural numbers through concrete examples.
(b) Repeat again and again: equivalent fractions are different numeral representations of the same number!
(c) Introduce improper fractions and mixed numbers in natural contexts as early as possible.
(d) Make extensive use of concrete visualizations; there is no gain in manipulating numbers and symbols before new concepts have been internalized.

28.1 Division of natural numbers

Consider the arithmetic expression:

$$2 \div 3.$$

By definition, its value is the size of a share when 2 objects, or 2 *wholes* are divided into 3 equal parts. Now that we have learned to express the *breaking of a whole*, we can use fractions to quantify the value of this quotient.

How much are 2 divided by 3? Here is one way to visualize this question:

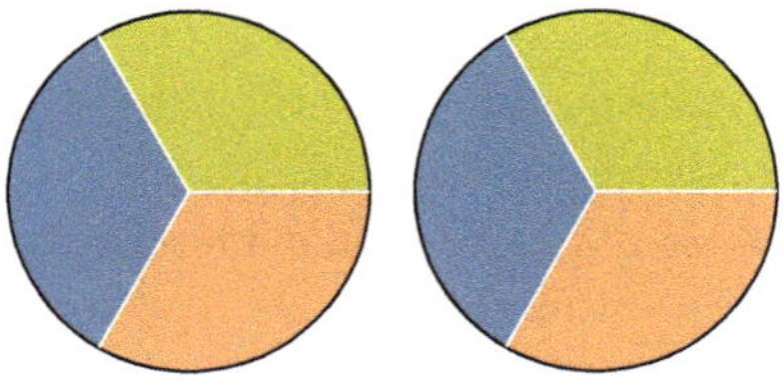

Recall that the first step, whenever using a concrete representation, is to define the whole. A priori, a whole could be anything; once a whole has been defined, all numbers mentioned in the discourse, whether natural numbers or fractions, must refer to that whole.

Probably the most convenient whole, when it comes to fractions, is a disc (also called a **pie**). What makes it so convenient is its symmetry, which allows its partition into any number of equal parts, or **sectors**. Moreover, given a sector, the whole that the sector is part of is evident. This is in contrast to, say, a segment. ONE-HALF of a segment still looks like a segment!

Consider then the above figure, which displays two discs, or two wholes. Each disc is partitioned into three equal sectors, each in a different color—one blue, one orange and one yellow. We have a total of 6 sectors of equal size, colored in 3 different colors. Next, rearrange the 6 sectors, such that sectors of the same color are grouped together:

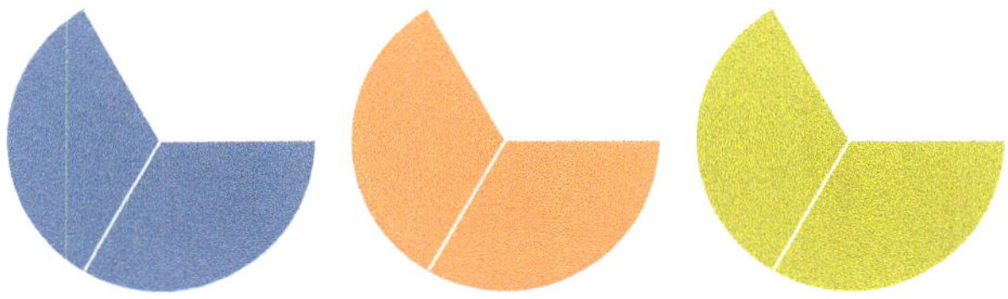

We have thus managed to partition the 2 discs into 3 equal parts. By definition, each part is the quotient obtained by dividing 2 discs into 3. Our task is to find the *value* of 2 divided by 3, i.e., we need to determine what fraction of a disc constitutes each of the 3 parts. The answer appears before us: each of the 3 parts is TWO-THIRDS of a disc, i.e., TWO-THIRDS of a whole. We conclude that

$$2 \div 3 = \frac{2}{3}.$$

A very important comment

In Chapter 27, we argued that the identity

$$1 \div 3 = \frac{1}{3}$$

is not the outcome of a calculation—it is a definition; the number ONE-THIRD is *defined* as the quotient of one whole divided by 3. Here the situation is totally different. The fraction TWO-THIRDS was defined (p. 174) as 2 times ONE-THIRD, and *not* as 2 divided by 3. Guided by a concrete model, we demonstrated that the value of the quotient $2 \div 3$ equals the fraction TWO-THIRDS. This is not a triviality!

The pie chart used to illustrate the arithmetic expression $2 \div 3$ can be used to illustrate any quotient of two natural numbers. Consider for example

$$4 \div 5.$$

Take 4 discs, partition each disc into 5 sectors, and color each sector by a different color:

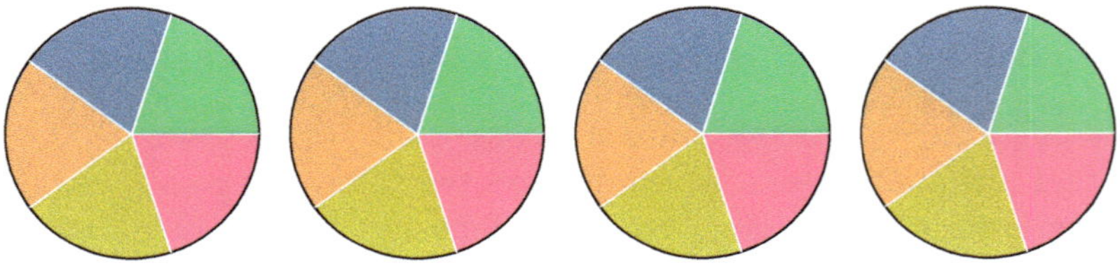

We have thus partitioned the 4 discs into 20 sectors. Rearrange these 20 sectors by color:

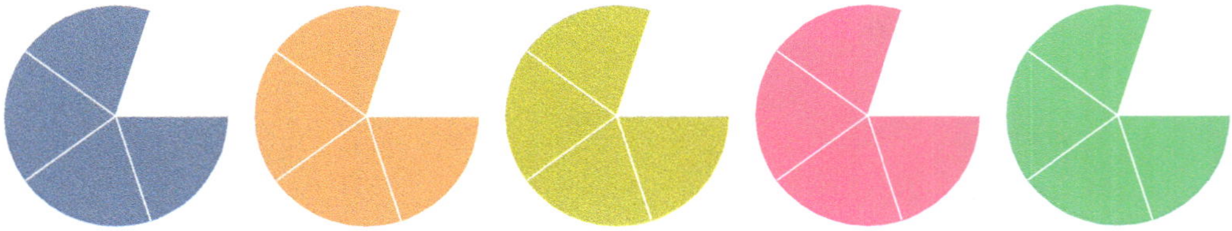

This arrangement shows that the size of a share when 4 discs are divided into 5 equal parts is FOUR-FIFTHS of a disc, namely,

$$4 \div 5 = \frac{4}{5}.$$

To generalize, let m and n be natural numbers. Their quotient is a fraction, whose denominator equals the divisor and whose numerator equals the dividend,

$$m \div n = \frac{m}{n}.$$

28.2 Equivalent fractions

Consider the word problem:

> Boaz cut a cake into 4 equal slices. Roni took 2 slices. What fraction of the cake did Roni take?

Solution: Each portion is ONE-QUARTER of the cake. Since Roni took two portions, she took TWO-QUARTERS of the cake.

An illustration of the solution is displayed below. The cake is modeled by a disc, and the part taken by Roni is colored blue:

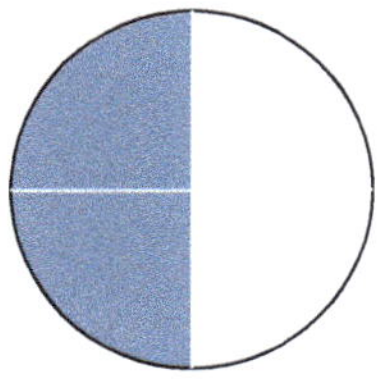

Examine the illustration: Exactly ONE-HALF of disc is colored blue; thus, we can also say that Roni got ONE-HALF of the cake.

Does this example suggest that TWO-QUARTERS and ONE-HALF are the same number? At a first glance, these two fractions may seem different. The fraction TWO-QUARTERS represents twice the size of a share when a whole is divided into 4 equal parts, whereas the fraction ONE-HALF represents the size of a share when a whole is divided into 2 equal parts. Since the cake in the above word problem was cut into 4 equals parts, should we conclude that the correct answer is that Roni took TWO-QUARTERS of the cake? Maybe we should say instead that she took ONE-HALF of the cake? Maybe both answers are correct?

In Volume 1, p. 5, we emphasized that the cardinality of a set does not depend on how this set is arranged. Likewise, the value of a part of a whole does not depend on how this part is arranged, or whether it is in one piece or not. As the illustration demonstrates, the fractions ONE-HALF and TWO-QUARTERS have the *exact same value*, i.e.,

$$\frac{1}{2} = \frac{2}{4}.$$

The mathematical terminology is that ONE-HALF and TWO-QUARTERS are different **fractional representations** of the same rational number, or that they are **equivalent fractions**.

Please note: from a semantic point of view, which relates to the actions of *cutting* and *taking apart*, TWO-QUARTERS may be a better description for the fraction of the cake taken by Roni than ONE-HALF. From a mathematical point of view, there is absolutely no distinction between those two fractions. The concept of a fraction is independent of the procedure used to create that fraction.

28.2.1 Equivalent fractions and multiplicative scaling

The equivalence, or equality between the fractions ONE-HALF and TWO-QUARTERS chimes in with two rules already learned. The first rule is the role of fractions in quantifying the outcome of the division of natural numbers. As seen on p. 184, the fraction TWO-QUARTERS equals the quotient $2 \div 4$, whereas the fraction ONE-HALF, by definition, equals the quotient $1 \div 2$. By the transitivity of equality, the equivalence between the fractions ONE-HALF and TWO-QUARTERS implies an equality between the quotients

$$1 \div 2 \qquad \text{and} \qquad 2 \div 4.$$

The second rule was mentioned in Volume 1, p. 106. Quotients vary when the dividend, or the divisor or both change. Specifically,

(a) Magnification of the dividend results in a magnification of the quotient by the same factor.
(b) Magnification of the divisor results in a reduction of the quotient by the same factor.

It follows that if both the dividend and the divisor are magnified by the same factor, the quotient does not change.

For example, the quotients

$$21 \div 3 \qquad \text{and} \qquad 42 \div 6$$

are equal, because in the passage from the first quotient to the second quotient, both dividend and divisor were magnified by a factor of 2. This rule has a very intuitive interpretation: if twice as many items are distributed into twice as many sets, the size of each share remains unchanged.

Viewed in this context, the equality between $1 \div 2$ and $2 \div 4$ results from the rules of multiplicative scaling:

$$1 \div 2 = \underbrace{(2 \times 1)}_{\text{doubling the dividend}} \div \underbrace{(2 \times 2)}_{\text{doubling the divisor}} = 2 \div 4.$$

28.2.2 Infinitude of equivalent fractions

Having established that ONE-HALF and TWO-QUARTERS are equivalent fractional representations of the same number, the next question is whether that same number has other equivalent representations.

The following figures give the answer:

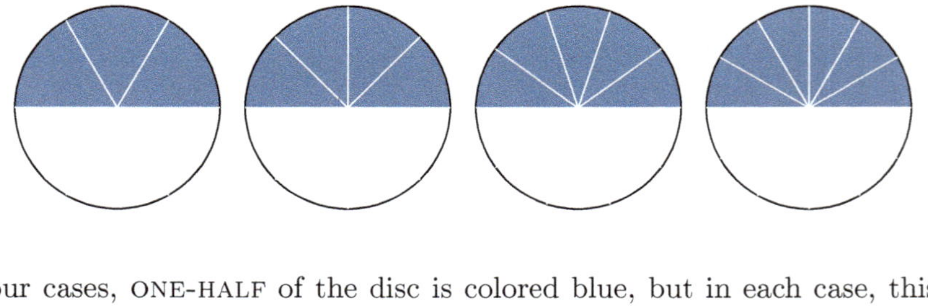

In all four cases, ONE-HALF of the disc is colored blue, but in each case, this half is partitioned differently; in each case the disc was cut into a different *even* number of equals sectors, and ONE-HALF of those sectors were colored. The parts that were colored correspond to the fractions

$$\frac{3}{6}, \quad \frac{4}{8}, \quad \frac{5}{10} \quad \text{and} \quad \frac{6}{12}.$$

Since those fractions quantify the same part of the disc, they are equivalent; they are all equal to ONE-HALF and to TWO-QUARTERS.

In this respect, there is nothing special about the fraction ONE-HALF. Consider the following figure,

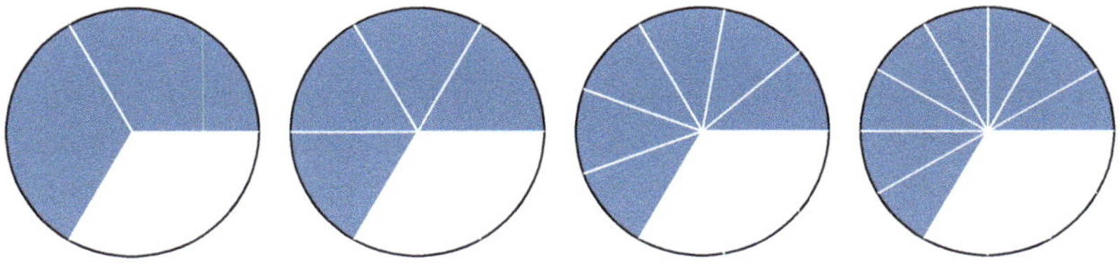

illustrating the equivalence of the fractions,

$$\frac{2}{3} = \frac{4}{6} = \frac{6}{9} = \frac{8}{12}.$$

We see that every fraction has an infinite sequence of equivalent fractional representations, obtained by subdividing the part it represents into smaller and smaller equal parts.

28.2.3 Natural numbers as fractions

Thus far, we have distinguished, within the realm of rational numbers, between natural numbers and fractions. In this section we will see that the distinction between the two is less categoric than it might seem at first.

Consider the figure:

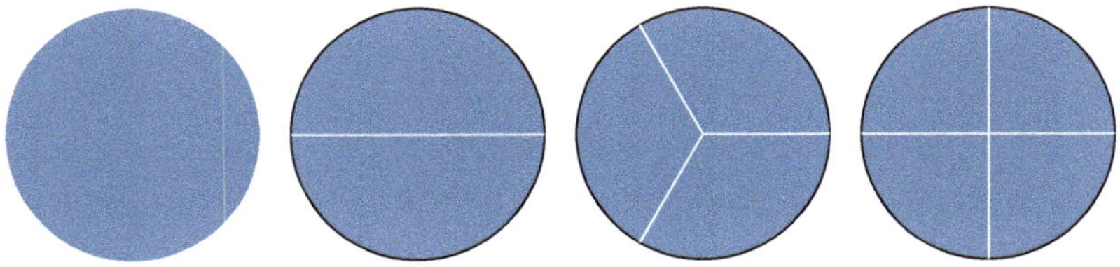

In each case, the part of the disc colored blue is the entire disc, which is represented by the number 1. However, in each case, the disc is partitioned into a different number of sectors, hence the equivalence illustrated in this figure is

$$1 = \frac{2}{2} = \frac{3}{3} = \frac{4}{4}.$$

Likewise, 2 wholes can be represented as FOUR-HALVES of a whole, SIX-THIRDS of a whole, etc. It turns out that natural numbers can also be represented by fractions.

Once again, this possibly surprising fact chimes in with prior knowledge. First, we learned in Volume 1, p. 112, that the quotient of every natural number divided by itself is one, namely,

$$1 = 2 \div 2 = 3 \div 3 = 4 \div 4 = \dots.$$

Second, we learned on p. 184 that for every natural number n, the fraction $^n/_n$ is the value of the quotient $n \div n$. Replacing $2 \div 2$ by TWO-HALVES, $3 \div 3$ by THREE-THIRDS, etc., we obtain the desired equivalence between the number 1 and fractions of the form $^n/_n$.

At this stage, you might be mind-boggled by the following question: if the quotient of every two natural numbers equals a fraction whose numerator is the dividend and whose denominator is the divisor, why did we have to work so hard in earlier grades? Consider for example, the arithmetic expression

$$24 \div 6.$$

It takes great effort to third graders memorizing that this quotient equals 4. Now that we have at our disposal fractions, we can represent the value of this quotient as

$$\frac{24}{6},$$

without having to calculate anything. Is anything wrong with it?

The answer to this provocative question is that the identity

$$24 \div 6 = \frac{24}{6}$$

is correct, however not always useful. When we evaluate a quotient, we want to express its value in an intelligible way. In most cases, a natural number is easier to grasp than an equivalent representation as a fraction. For purposes of clarification, suppose that this division arises from the word problem:

> 24 students were equally partitioned into 6 study groups. How many students are in each group?

Representing the solution as TWENTY-FOUR-SIXTHS suggests that each group comprises 24 human parts, each being ONE-SIXTH of a whole person. This answer is correct, but quite pointless.

Fractions with denominator 1

Thus far, we have always considered fractions with denominators of 2 or more. What about the denominator of 1? If we follow the definition of unit fractions, the fraction

$$\frac{1}{1}$$

quantifies the part of a whole when that whole is partitioned into 1 part (i.e., not partitioned). We deduce that the fraction $^1/_1$ equals 1, which is also consistent with $^1/_1$ being equal to the quotient $1 \div 1$.

Likewise, we may consider repeated addition of $^1/_1$, e.g.,

$$\frac{5}{1} = 5 \times \frac{1}{1} = 5.$$

The representation of natural numbers as fractions with denominators of 1 may seem pointless. Nevertheless, it is sometimes useful, for example, when we want to devise an algorithm applicable to natural numbers and fractions alike.

28.2.4 Expansion and reduction

Every rational number, whether a natural number or a fraction, has an infinite number of equivalent fractional representations. As we will see in the subsequent chapters, representing the same number by different fractional representations is of immense practical importance.

The transition between equivalent representations involves either the magnification or the reduction of both denominator and numerator by the same natural numbers. Magnifying both the denominator and the numerator by the same natural number is called **fraction expansion**. Reducing both denominator and numerator by the same natural number is called **fraction reduction**. This section elaborates both procedures.

28.2.4.1 Expansion

An expanded representation of a fraction is obtained by magnifying both denominator and numerator by a natural number—in principle, *any* natural number. Consider, for example, the fraction

$$\frac{5}{7}.$$

Magnifying both denominator and numerator by a factor of 3, we obtain the equivalent fraction

$$\frac{3 \times 5}{3 \times 7} = \frac{15}{21}.$$

There can be no over-repetition: FIVE-SEVENTHS and FIFTEEN-OVER-TWENTY-ONE are different fractional representations of the *same* number. They are mathematically indistinguishable. Expanding the fraction FIVE-SEVENTHS by a factor of 3 is not the same as magnifying it by a factor of 3, despite the semantic similarity between the verbs *expand* and *magnify* in spoken language. Expanding a fraction does not change its value.

We next provide a concrete illustration to the expansion of FIVE-SEVENTHS by a factor of 3; for the sake of variation, let's take a rectangle as a concrete representation of a whole. In the following figure, FIVE-SEVENTHS of the rectangle have been colored blue:

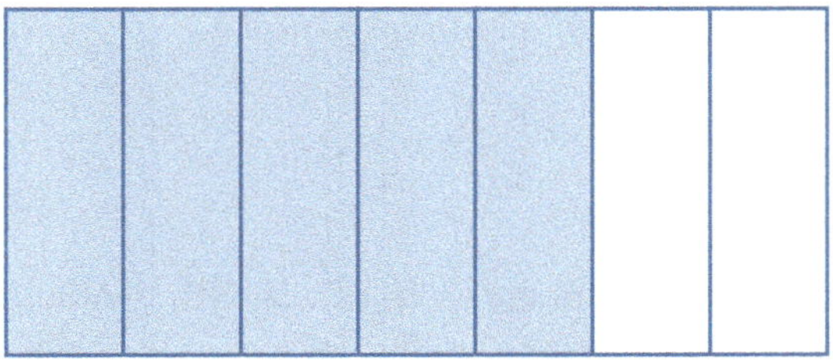

To do so, we partitioned the rectangle into 7 equal parts by using vertical separators, each being ONE-SEVENTH of the rectangle, and colored 5 out of the 7 parts. Take now each ONE-SEVENTH of the rectangle, and partition it into 3, as follows:

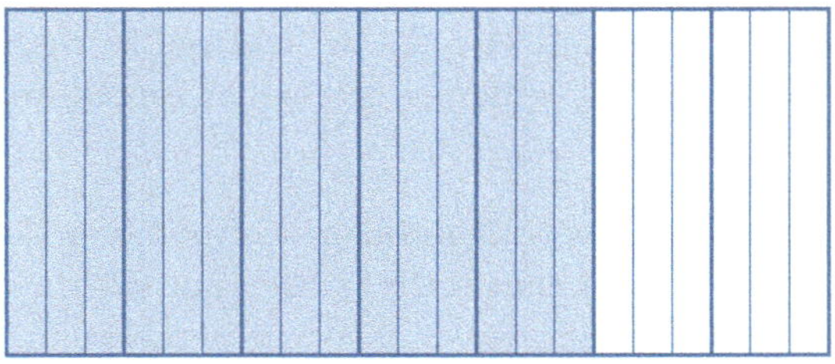

Now, the rectangle is partitioned into $3 \times 7 = 21$ parts, and the number of blue parts tripled as well; there are $3 \times 5 = 15$ of them. Thus, the fraction of the rectangle that is blue is FIFTEEN-OVER-TWENTY-ONE, which is the same fractional part as FIVE-SEVENTHS.

28.2.4.2 Reduction

A fraction is reduced by dividing its denominator and its numerator by the same natural number. Since both denominator and numerator must be natural numbers, reduction can only be carried out if this division yields integer quotients. In particular, there exist fractions that cannot be reduced; such fractions are said to be **irreducible**.

In order for a fraction to be reducible, the numerator and the denominator must have a **common divisor**. Consider, for example, the fraction

$$\frac{24}{42}.$$

Its numerator, 24, and its denominator, 42, have a common divisor, 2. Reducing the fraction by a factor of 2, we obtain the equivalent fraction

$$\frac{24 \div 2}{42 \div 2} = \frac{12}{21}.$$

This latter fraction is further reducible, as 12 and 21 have a common divisor, 3. Reducing the fraction by a factor of 3, we obtain the equivalent fraction

$$\frac{12 \div 3}{21 \div 3} = \frac{4}{7}.$$

The latter fraction is irreducible, as 4 and 7 have no common divisor.

To obtain the irreducible equivalent representation of a given fraction, we reduce it by the **greatest common divisor** of the denominator and the numerator (see Chapter 24). In the above example, 6 is the greatest common divisor of 24 and 42, hence TWENTY-FOUR-OVER-FORTY-TWO can be brought into an irreducible equivalent form by reducing it by a factor of 6,

$$\frac{24}{42} = \frac{24 \div 6}{42 \div 6} = \frac{4}{7}.$$

28.3 Improper fractions and mixed numbers

28.3.1 Improper fractions

Consider the word problem:

> Each day, Tal's hamster eats ONE-THIRD of an apple. What fraction of an apple will it eat in 4 days?

The answer is easily obtained by repeated addition,

$$\frac{1}{3} + \frac{1}{3} + \frac{1}{3} + \frac{1}{3} = \frac{4}{3}.$$

Remember that this equality is the definition of FOUR-THIRDS rather than the outcome of a calculation.

If we have been accustomed, so far, to viewing fractions as expressing quantities that are less than a whole, this is not the case here: THREE-THIRDS equal a whole, hence FOUR-THIRDS are more than a whole. Parts of a whole may accumulate together into a quantity that is more than a whole, as in the present example. Since for every natural number n,

$$\frac{n}{n} = 1,$$

it follows that fractions representing quantities greater than a whole are characterized by their numerator being greater than their denominator. Fractions representing quantities greater than a whole are called **improper fractions**; in contrast, fractions representing quantities less than a whole are called **proper fractions**. It should be emphasized that there is nothing "improper" in improper fractions; they are valid fractional representations of rational numbers.

28.3.2 Mixed numbers

A concrete illustration of the improper fraction FOUR-THIRDS is depicted below. As in previous examples, the whole is represented by a disc:

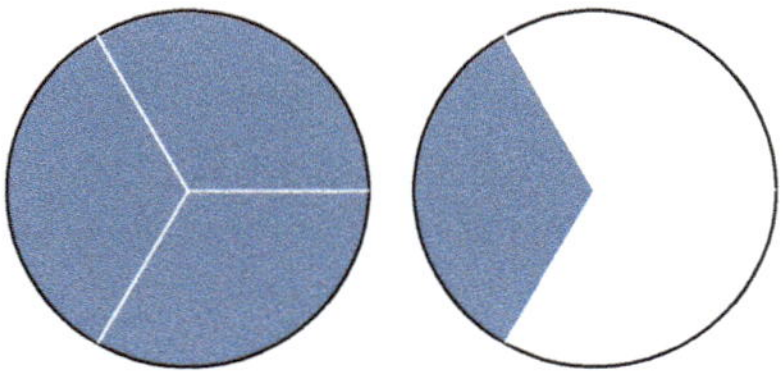

What fraction of a disc is colored blue? The most natural answer would be that the blue part comprises a whole plus a ONE-THIRD of a whole, namely.

$$1 + \frac{1}{3}.$$

The sum of a natural number and a fraction is commonly denoted by removing the plus sign, and writing the natural number and the fraction side-to-side as follows:

$$1\frac{1}{3}.$$

In words: ONE-AND-ONE-THIRD.

The combination of a natural number and a fraction is called a **mixed number**. A mixed number is an alternative representation for improper fractions, separating the integer part from the fractional part. Thus,

$$\frac{4}{3} \qquad \text{and} \qquad 1\frac{1}{3}$$

represent the same number, the first in the form of an improper fraction and the second in the form of a mixed number. At this stage, you should no longer be disturbed by the same number having multiple numeral representations.

28.3.3 From improper fractions to mixed numbers

The ability to switch from an improper fraction representation to an equivalent mixed number representation, and vice versa, is a useful skill. This is the subject of this and the next section.

Consider the fraction

$$\frac{27}{4}.$$

As we already know, it is the value of the quotient

$$27 \div 4.$$

In Chapter 22 we learned to express the outcome of division as a sum of a quotient and a remainder. In this case,

$$27 \div 4 = 6 + \mathrm{R}\,(3 \div 4).$$

In words: the outcome of dividing 27 wholes into 4 equals parts is 6 wholes, plus a remainder of 3 wholes that ought to be divided into 4 equals parts. Now, however, we can express the quotient $3 \div 4$ as a fraction—THREE-QUARTERS. Hence,

$$\frac{27}{4} = 27 \div 4 = 6 + \frac{3}{4} = 6\frac{3}{4}.$$

Note how the remainder $\mathrm{R}\,(3 \div 4)$ is replaced by the fraction THREE-QUARTERS. Anyone who has learned to denote remainders using the notation proposed in Chapter 22 has a ready-to-use method for converting improper fractions into mixed numbers.

28.3.4 From mixed numbers to improper fractions

Consider the mixed number

$$5\frac{6}{7}.$$

Our goal is to rewrite it as an equivalent improper fraction.

We proceed in three stages:

(a) A mixed number is a short-hand notation for the sum of a natural number and a fraction:

$$5\frac{6}{7} = 5 + \frac{6}{7}.$$

(b) The natural number 5 represents a repeated addition of ones, whereas the fraction SIX-SEVENTHS represents a repeated addition of ONE-SEVENTHs. We next use the fact that whole numbers can also be represented as fractions. Since the denominator of the fractional part of the mixed number is 7, we express the whole as SEVEN-SEVENTHS, so that

$$5 + \frac{6}{7} = 5 \times \frac{7}{7} + \frac{6}{7}.$$

(c) In the last stage, we evaluate the number of sevenths in this expression: there are 5 times SEVEN-SEVENTHS, plus SIX-SEVENTHS, making together FORTY-ONE-SEVENTHS, i.e.,

$$5\frac{6}{7} = \frac{41}{7}.$$

Another approach would be to resort again to division with a remainder, this time going in the opposite direction. We are looking for a natural number satisfying the following equation,

$$\boxed{?} \div 7 = 5 + \mathrm{R}\,(6 \div 7)$$

(see the Problems section in Chapter 22). This number is 41. Using twice the identity $m \div n = {}^{m}/{}_{n}$, we obtain that

$$\frac{41}{7} = 41 \div 7 = 5 + \mathrm{R}\,(6 \div 7) = 5 + \frac{6}{7}.$$

Mathematical problems

Problem 28.1. Propose a concrete illustration of the quotient

$$3 \div 4$$

using a metaphor of 3 bananas divided equally among 4 children. Is there more than one possible illustration?

Problem 28.2. Explain, using a concrete visualization, why

$$8 \div 9 = 8 \times \frac{1}{9}.$$

Problem 28.3. What fraction of an hour are 6 minutes? 20 minutes? 65 minutes?

Problem 28.4. Activate a metronome at a pace of 30 beats per second. You can find free online metronomes. Try to tap with both hands simultaneously: 2 taps per beat with your right hand and 3 taps per beat with your left hand. Unless you are a trained musician, you might find this task quite challenging. Try to overcome the difficulty by partitioning every beat into 6.

Pedagogical problems

Problem 28.5. A group of children was asked to evaluate the expression

$$5 \div 2.$$

Gili replied: It is not divisible.
Alon replied: 2 and a remainder of 1.
Tal replied: TWO-AND-A-HALF.
Anat replied: FIVE-HALVES.

(a) Sort the answers into correct and incorrect.
(b) Is there one answer that is more correct than the others?
(c) Find a concrete visualization for each answer.

Problem 28.6. Teacher Kalagian asked her class to evaluate the quotient,

$$2 \div 3.$$

Sivan replied: The answer is a third. The whole is 2, and it is divided into 3 equal parts. Hence, each part is ONE-THIRD.

(a) What is the misconception underlying Sivan's wrong answer?
(b) How would you correct her?

Chapter 29

Fraction Comparison

In the last two chapters, we became acquainted with the set of **rational numbers**, which comprises natural numbers and fractions. When a mathematical system is extended, we usually wish to ascertain that key properties of the system prior to the extension are preserved after the extension. In the case of number systems, properties that ought to be preserved are the commutative, associative and distributive properties of addition and multiplication. In the subsequent chapters we will see that these properties are indeed preserved by the set of rational numbers.

Another key property of the number system, which ought to be preserved after including fractions, is the **order relation**—the ability to compare every combination of two numbers using a **less than/equal to/greater than** relation (see Volume 1, p. 6). One may even say that the ability to determine which of two quantities is greater was the primary cause for inventing numbers. Comparison between natural numbers was examined in detail in Chapter 1. This section considers comparison between fractions.

Hurdles:

(a) Children often do not relate the procedure of fraction comparison to the notions of *greater than* and *less than*.
(b) Children often associate a larger denominator with a larger fraction.
(c) Expanding two fractions to obtain a common denominator is a multi-step procedure.

Remedies:

(a) Use visual aids to relate fractions to the size of objects, *and* to the cardinality of subsets.
(b) Relate fraction comparison to a comparison of quotients of whole numbers.

(c) Use concrete models to demonstrate the procedure of obtaining a common denominator; emphasize that expansion and reduction change the representation of a fraction, but not its value.

Opportunities:

(a) Exploit the subject of fraction comparison to consolidate children's perception of the fraction concept.
(b) Encourage children to find creative way of comparing fractions.

29.1 Order relations and inclusion

The order relation between natural numbers is based on the following principle: Two sets, Set A and Set B, are said to be of **equal size** if there exists an **exact matching** among their elements: one can find a one-to-one correspondence between objects from one set to objects of the other set, so that there remain no unmatched objects in either set. Likewise, Set A is said to be smaller, or contain fewer objects than Set B, if there exists an exact matching between *all* the elements of Set A and *part* of the elements of Set B. The following figure depicts the latter relation:

The natural numbers inherit the order relations among sets: the number 3 is said to be less than the number 5, because a set comprising 3 objects can be matched exactly to part of a set comprising 5 objects, or because a set of 5 objects includes a subset of 3 objects. In other words, the order relation among numbers is induced by an **inclusion relation** among sets.

When it comes to rational numbers—natural numbers and fractions alike—we define an order relation based on the exact same principle:

Definition 29.1. A rational number a is said to be less than a rational number b, if a wholes (or a fraction a of a whole) are contained in b wholes (or a fraction b of a whole).

For example,

$$\frac{1}{4} < \frac{1}{2}$$

because we can make a one-to-one matching between ONE-QUARTER of a disc and *part* of ONE-HALF of a disc, as depicted below:

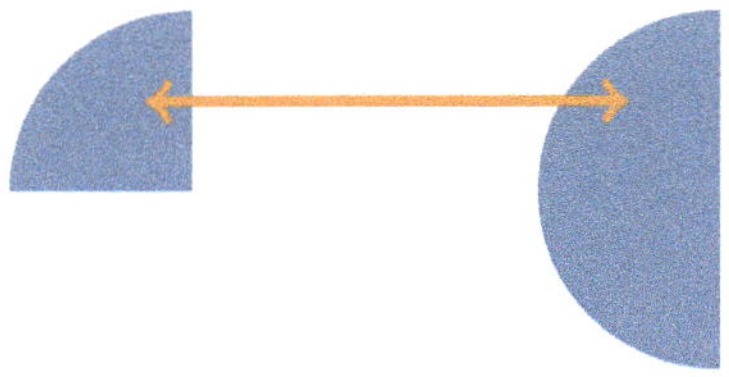

A reminder: the abstract noun *object* can be replaced by a concrete noun, such as *apple* or *elephant*. The important thing to remember is that once the *whole*, *unit* or ONE has been assigned a concrete semantic meaning, one must continue to stick to this concrete whole. To drive things to absurdity, one cannot claim that ONE-QUARTER is greater than 10 because ONE-QUARTER of an hour is 15 minutes, hence greater than 10 minutes. Either the whole is an hour, or it is a minute. The choice is arbitrary (usually driven by practical consideration), but must be explicit and unambiguous. Once the determination of the whole has been made, all numbers must refer to that whole; numbers representing different wholes cannot be compared.

29.2 Fraction comparison

In this section we will encounter a variety of ways for determining which of two given rational numbers is greater than the other. Since we are already familiar with the comparison of natural numbers, we will devote this section exclusively to the comparison of fractions.

There exists a systematic approach, based on finding a **common denominator**, for comparing every two fractions. However, there are excellent reasons for deferring its presentation to a later stage, and opening this section with a variety of alternative methods, which are less structured than the systematic approach. The reasons are:

(a) The systematic approach is not always the most efficient. You shouldn't look for a common denominator to determine that THREE-QUARTERS are greater than ONE-THIRD.

(b) Mastery of different methods of comparison deepens one's understanding of the fraction concept.

(c) Self-discovery of alternative methods is at the heart of mathematical endeavor.

29.2.1 Comparison using concrete models

In the transition from natural numbers to fractions, there is a tendency among educators to resort to abstraction and procedural methods too soon. I believe that part of the "fraction trauma" could be avoided by providing children with sufficient concrete, hands-on experience throughout the study of fractions. If fact, even adults, e.g., teachers and parents, could benefit from getting acquainted with varied representations of fractions, whether using illustrations, concrete objects or even thought experiments. In this section we review several methods for comparing fractions using concrete representations.

29.2.1.1 Comparing parts of a whole

Which fractions is greater,

$$\frac{3}{5} \quad \text{or} \quad \frac{2}{3}?$$

One way to transform this abstract question into a more tangible one, is to rephrase it as follows:

> Which fraction of a whole is greater: THREE-FIFTHS of that whole or TWO-THIRDS of that whole?

This subtle rephrasing has the virtue of putting forward the fact that both fractions represent parts of the same whole. As iterated before: there are no fractions without a *unit* or a ONE. Concretization starts with the choice of a whole.

In most examples that we saw so far, the whole was illustrated using a disc. Since abstraction abilities grow from a continual exposure to varied concrete examples, we will use here a different whole. In ancient Greece, the canonical unit was a rod, whose length was attributed the value 1. Fractions were represented by parts of that rod (see p.177). The fraction THREE-FIFTHS is represented by a rod obtained by concatenating 3 rod segments, each contained 5 times in the unit rod. Similarly, the fraction TWO-THIRDS is represented by a rod obtained by concatenating 2 rod segments, each contained 3 times in the unit rod; see figure:

Please note that in order to generate a rod that is TWO-THIRDS of the unit rod, one has to first cut the unit rod into 3 segments of equal length and then concatenate 2 of those segments end-to-end. Similarly, in order to generate a rod that is THREE-FIFTHS of the unit rod, one cuts the unit rod into 5 segments of equal length and then concatenates 3 of those segments end-to-end. On the longer run, such a procedure is impractical, but for starters, it offers a concrete experience of what fractions are and how to compare their magnitude. Moreover, many scholar kits offer **fraction bars**, which are collections of unit rods pre-partitioned into sub-segments of equal length.

It is not obvious, at first sight, which of the two rods—THREE-FIFTHS of a unit rod or TWO-THIRDS of a unit rod—is the longest (this is not a coincidence; I chose fractions differing by little). To find which fraction is greater, we need to juxtapose the two rods and see which is longer:

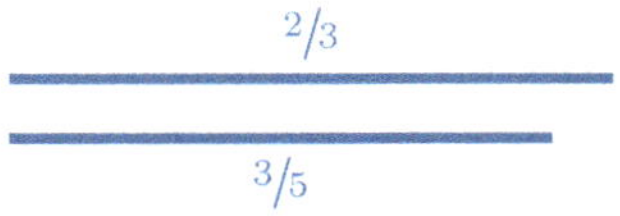

We conclude that,

$$\frac{3}{5} < \frac{2}{3}.$$

It is instructive to examine another concrete illustration of this example. Instead of using a "bare" disc, we will use the face of a clock, and identify the whole with the area swept by the minutes hand in an hour.

The fraction THREE-FITHS is represented by 3 times ONE-FIFTH of the area swept by the minutes dial in an hour. ONE-FIFTH of an hour is the time interval obtained when dividing the hour into 5 equal intervals. Since an hour is 60 minutes, ONE-FIFTH of an hour is $60 \div 5 = 12$ minutes, and THREE-FIFTHS of an hour are therefore $3 \times 12 = 36$ minutes.

Similarly, the fraction TWO-THIRDS is represented by 2 times ONE-THIRD of the area swept by the minutes dial in an hour. ONE-THIRD of an hour is the time interval obtained when dividing the hour into 3 equal intervals, i.e., $60 \div 3 = 20$ minutes, and TWO-THIRDS of an hour are $2 \times 20 = 40$ minutes.

The areas swept by the minutes dial in THREE-FIFTHS of an hour and TWO-THIRDS of an hour are depicted below. This illustration proves once again the inequality

$$\frac{3}{5} < \frac{2}{3}.$$

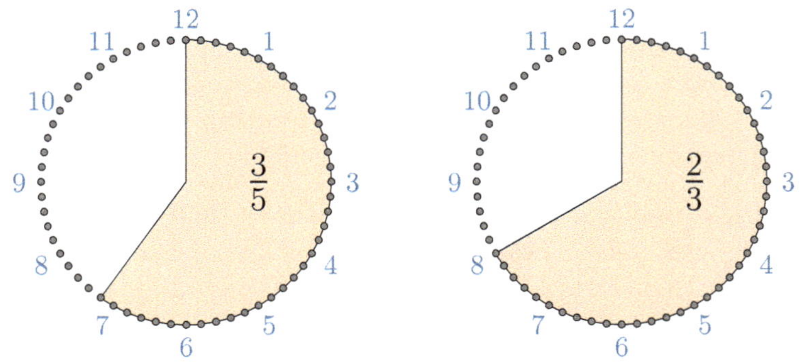

I wish to draw your attention to the central role that a 60-minute hour plays in this analysis. Since 60 is divisible by both 5 and 3, it was relatively straightforward to establish how long THREE-FIFTHS of an hour and TWO-THIRDS of an hour are. Keep this in mind when we learn to compare fractions using a **common denominator**. Also, convince yourself that this illustration is not as effective when comparing, say, THREE-EIGHTHS and FOUR-ELEVENTHS.

29.2.1.2 Comparing quotients of natural numbers

Consider the same question again: which fraction is greater,

$$\frac{3}{5} \quad \text{or} \quad \frac{2}{3}?$$

In this section we will use the fact that the fraction m/n is the quotient when dividing m wholes into n equal shares (see p. 184). Thus, we may rephrase the question as follows:

> Which is greater: the size of a share when partitioning 3 wholes into 5 equals parts, or the size of a share when partitioning 2 wholes into 3 equals parts.

This problem reformulation may not seem easier to grasp than the original one (it is of course equivalent to it). But now, let's concoct a corresponding word problem:

> What would you prefer: sharing 3 cookies among 5 children, or sharing 2 cookies among 3 children?

You might be surprised to discover how mathematical thinking sharpens when cookies are at stake!

One could solve this question using, for example, the **chunking** strategy, introduced on p. 114 of Volume 1, and also on p. 45 in this volume. Start by distributing ONE-HALF of a cookie to each child. The 5 children who are sharing 3 cookies will get ONE-HALF of a cookie each, leaving a remainder of ONE-HALF of a cookie:

The 3 children who are sharing 2 cookies will get ONE-HALF of a cookie each, leaving a remainder of ONE-HALF of a cookie as well:

Thus, each child in both groups has gotten the same amount of cookies so far, and each group has a remainder of ONE-HALF of a cookie. However, the children in the fivesome have to share the remaining ONE-HALF of a cookie among 5, whereas the children in the threesome have to share the remaining ONE-HALF of a cookie among 3. Obviously, it is more beneficial to be part of the smaller group, hence,

$$\frac{3}{5} < \frac{2}{3}.$$

29.2.2 Fractions with like denominators

Which fraction is greater,

$$\frac{2}{5} \quad \text{or} \quad \frac{3}{5}?$$

The answer might seem obvious, but an explicit analysis can lead to rewarding insights.

As we did before, let's rephrase the problem in a slightly different way:

> Which fraction of a whole is greater: TWO-FIFTHS of that whole or THREE-FIFTHS of that whole?

ONE-FIFTH of a whole is the size of a share obtained when dividing that whole into 5 equal parts. General fractions are multiples of a unit fraction having the same denominator. With this in mind, TWO-FIFTHS of the whole is 2 times ONE-FIFTH of that whole, and THREE-FIFTHS of the whole is 3 times ONE-FIFTH of that whole. Since THREE-FIFTHS of a whole *include* TWO-FIFTHS of that whole,

$$\frac{2}{5} < \frac{3}{5}.$$

Please note: THREE-FIFTHS is greater than TWO-FIFTHS for the same reason that 3 is greater than 2. In fact, we could have rephrased the question as follows:

> Which is greater: 2 fifths-of-a-whole or 3 fifths-of-a-whole?

What we have done here, implicitly, is to define a new whole, which is ONE-FIFTH of the original whole. We have thus reduced the problem into comparing natural numbers—2 and 3. Modifying the whole is legitimate, as long as *all* the numbers refer to that new whole. You can't compare two numbers that refer to different wholes.

The same considerations can be applied to compare any two fractions having the same denominator. If k, m and n are natural numbers, then

$$\frac{k}{n} < \frac{m}{n} \qquad \text{if and only if} \qquad k < m,$$

because

$$\frac{k}{n} = \underbrace{\frac{1}{n} + \frac{1}{n} + \cdots + \frac{1}{n}}_{k \text{ times}} \qquad \text{and} \qquad \frac{m}{n} = \underbrace{\frac{1}{n} + \frac{1}{n} + \cdots + \frac{1}{n}}_{m \text{ times}}.$$

29.2.3 Fractions with like numerators

Which fraction is greater,

$$\frac{1}{5} \quad \text{or} \quad \frac{1}{7}?$$

or slightly rephrased,

> Which fraction of a whole is greater: ONE-FIFTH of that whole or ONE-SEVENTH of that whole?

ONE-FIFTH of a whole is the size of a share obtained when that whole is divided into 5 equals parts. ONE-SEVENTH of a whole is the size of a share when that whole is divided into 7 equals parts. In Volume 1, p. 107, we learned a fundamental property of division:

the greater the divisor, the less is the quotient. In simpler terms, if you divide something into more parts, each part is smaller. Thus,

$$\frac{1}{5} > \frac{1}{7}.$$

This inequality may be confusing if one mistakingly assigns a larger value to a fraction composed of larger numbers.

The same considerations can be generalized to any two unit fractions. If m and n are natural numbers, then

$$\frac{1}{m} < \frac{1}{n} \qquad \text{if and only if} \qquad m > n,$$

because a greater denominator amounts to a greater divisor.

Next, let's find which fraction is greater,

$$\frac{5}{7} \quad \text{or} \quad \frac{5}{9}?$$

Invoking the definition of general fractions,

$$\frac{5}{7} = \frac{1}{7} + \frac{1}{7} + \frac{1}{7} + \frac{1}{7} + \frac{1}{7} \qquad \text{and} \qquad \frac{5}{9} = \frac{1}{9} + \frac{1}{9} + \frac{1}{9} + \frac{1}{9} + \frac{1}{9}.$$

In both cases, the fraction equals 5 times a unit fraction—ONE-SEVENTH in the first case and ONE-NINTH in the second case. Since ONE-SEVENTH is greater than ONE-NINTH,

$$\frac{5}{7} > \frac{5}{9}.$$

Once again, the same considerations can be generalized to comparing any two fractions that have the same numerator. If k, m and n are natural numbers, then

$$\frac{k}{m} < \frac{k}{n} \qquad \text{if and only if} \qquad m > n.$$

Explanation:

$$\frac{k}{m} = \underbrace{\frac{1}{m} + \frac{1}{m} + \cdots + \frac{1}{m}}_{k \text{ times}} \qquad \text{and} \qquad \frac{k}{n} = \underbrace{\frac{1}{n} + \frac{1}{n} + \cdots + \frac{1}{n}}_{k \text{ times}},$$

and

$$\frac{1}{m} < \frac{1}{n} \qquad \text{if and only if} \qquad m > n.$$

29.2.4 Finding a common denominator

29.2.4.1 The concept

We have seen two instances in which fraction comparison happens to be straightforward: fractions with like denominators and fractions with like numerators. Let's return to the problem with which we opened this section: which fraction is greater,

$$\frac{3}{5} \quad \text{or} \quad \frac{2}{3}?$$

If we examine the numerators, we might be tempted to say that THREE-FIFTHS are larger. If we examine the denominators instead, we might be tempted to say that TWO-THIRDS are larger. Clearly, when making comparisons, one cannot ignore one of the two numbers composing a fraction.

When no simple criterion is visible, the key to fraction comparison is based on the fact that every fraction has infinitely many equivalent representations (p. 187). In particular, we can show that every two fractions have equivalent representations sharing the same denominator. The strategy is to expand (or reduce) both fractions, such to obtain equivalent representations having a common denominator, and then compare the numerators (p. 201). Remember that expanding a fraction does not change its value—it only changes its representation!

Specifically, the fraction THREE-FIFTHS has the following equivalent representations,

$$\frac{3}{5} = \frac{6}{10} = \frac{9}{15} = \frac{12}{20} = \frac{15}{25} = \cdots.$$

The fraction TWO-THIRDS has the following equivalent representations,

$$\frac{2}{3} = \frac{4}{6} = \frac{6}{9} = \frac{8}{12} = \frac{10}{15} = \frac{12}{18} = \cdots.$$

Examining both sequences, we realize that THREE-FIFTHS and TWO-THIRDS have equivalent representations with denominator 15. Comparing the numerators, we deduce that

$$\left(\frac{9}{15} =\right) \quad \frac{3}{5} < \frac{2}{3} \quad \left(= \frac{10}{15}\right).$$

This method for comparing two fractions is always feasible; it is always possible to expand two fractions such that their denominators coincide, because every two natural numbers have a common multiplier. The key to applying this method successfully and efficiently, is learning to find this common denominator quickly.

29.2.4.2 Finding a common denominator

The expansion of a fraction requires magnifying its denominator and numerator by the same factor. We are looking for *two* expansion factors, one for each fraction, such that the denominators end up being equal. Since the denominators of equivalent expanded fractions are multiples of the original denominator, we are looking, in fact, for a **common multiple** of the two denominators.

The practice of finding common multiples was exposed on p. 109. Every two natural numbers, m, and n, have infinitely many common multiples, among which there is a minimal one, denoted by $\text{LCM}(m, n)$.

We next examine alternative means of expanding fractions to obtain a common denominator. Our exposition is based on the principles described in Chapter 24. The running example in this section will be:

Compare the fractions TEN-FOURTEENTHS and EIGHTEEN-OVER-TWENTY-ONE.

Method 1: Product of denominators

Since the product of every two numbers is their joint multiple, we may obtain a common denominator by expanding each fraction by the denominator of the other fraction:

$$\frac{10}{14} = \frac{21 \times 10}{21 \times 14} = \frac{210}{294} \qquad \text{and} \qquad \frac{18}{21} = \frac{14 \times 18}{14 \times 21} = \frac{252}{294}.$$

Comparing numerators,

$$\left(\frac{210}{294} =\right) \quad \frac{10}{14} < \frac{18}{21} \quad \left(= \frac{252}{294}\right).$$

Method 2: Least common multiple

Every pair of fractions has infinitely many common denominators. Since larger denominators result in longer calculations, it is desirable to find the smallest common denominator, which is the least common multiple of the two denominators.

In Chapter 24 we learned various methods for finding the least common multiple of every two numbers. The most elegant method is based on prime number factorization. The prime factorizations of 14 and 21 are

$$\mathcal{P}(14) = \{2, 7\} \qquad \text{and} \qquad \mathcal{P}(21) = \{3, 7\}.$$

The prime factorization of their least common multiple is

$$\mathcal{P}(\text{LCM}(14, 21)) = \{2, 3, 7\},$$

i.e., the least common multiple is

$$\text{LCM}(14, 21) = 2 \times 3 \times 7 = 42.$$

Expanding the two fractions accordingly,

$$\frac{10}{14} = \frac{3 \times 10}{3 \times 14} = \frac{30}{42} \qquad \text{and} \qquad \frac{18}{21} = \frac{2 \times 18}{2 \times 21} = \frac{36}{42}.$$

Comparing numerators,

$$\left(\frac{30}{42} =\right) \quad \frac{10}{14} < \frac{18}{21} \quad \left(= \frac{36}{42}\right).$$

Method 3: First reduce

Both fractions TEN-FOURTEENTHS and EIGHTEEN-OVER-TWENTY-ONE are reducible. Often, the best practice for finding a common denominator is to first bring both fractions to irreducible form.

One could try reducing a fraction by trial and error, looking for common divisors of its numerator and its denominator. As described on p. 191, an irreducible representation is obtained when the reduction factor is the greatest common divisor of the numerator and the denominator. Since the prime number factorization of 10 and 14 is,

$$\mathcal{P}(10) = \{2, 5\} \qquad \text{and} \qquad \mathcal{P}(14) = \{2, 7\},$$

then, their greatest common divisor is 2. Since the prime number factorization of 18 and 21 is

$$\mathcal{P}(18) = \{3, 6\} \qquad \text{and} \qquad \mathcal{P}(21) = \{3, 7\},$$

then their greatest common divisor is 3. Irreducible representations are obtained by reducing TEN-FOURTEENTHS by a factor of 2 and reducing EIGHTEEN-OVER-TWENTY-ONE by a factor of 3,

$$\frac{10}{14} = \frac{10 \div 2}{14 \div 2} = \frac{5}{7} \qquad \text{and} \qquad \frac{18}{21} = \frac{18 \div 3}{21 \div 3} = \frac{6}{7}.$$

At it turns out, we obtained irreducible representations having a common denominator, therefore

$$\left(\frac{5}{7} =\right) \quad \frac{10}{14} < \frac{18}{21} \quad \left(= \frac{6}{7}\right).$$

In general, however, reduction will not result in common denominators, so an application of either Method 1 or Method 2 is necessary thereafter.

29.2.5 Finding a common numerator

Which fraction is greater,

$$\frac{4}{47} \quad \text{or} \quad \frac{8}{85}?$$

Both fractions are irreducible. Moreover, the least common multiple of 47 and 85 is 47×85, hence a comparison based on a common denominator is lengthy and prone to errors. A simpler approach, in this case, would be to expand the fractions to obtain a **common numerator**, and compare denominators (recall: the larger the denominator, the less is the fraction).

Expanding the first fraction by a factor of 2,

$$\frac{4}{47} = \frac{2 \times 4}{2 \times 47} = \frac{8}{94},$$

we deduce that

$$\left(\frac{8}{94} =\right) \quad \frac{4}{47} < \frac{8}{85}.$$

Surprisingly, this method, which often saves considerable work, is rarely taught in school. The reason for preferring common denominators to common numerators is that the former is useful not only for comparing fractions, but also for adding and subtracting them (see next chapter). Nevertheless, I strongly encourage adding the common numerator strategy to the children's toolbox. A creative mathematician always looks for the most elegant solution to a given problem. This is an example of being "lazy" in the positive sense of the word.

29.2.6 Comparison using an intermediary

Which fraction is greater,

$$\frac{81}{197} \quad \text{or} \quad \frac{71}{211}?$$

Neither strategy, either seeking a common denominator or seeking a common numerator, is appealing. We note, however, that the first fraction has both a greater numerator and a smaller denominator, hence it must be larger.

To convince ourselves that this is indeed the case, without performing any calculation, we form an **intermediary** fraction using the numerator of one fraction and the denominator of the other fraction. We compare this intermediary fraction to both original fractions. Comparing numerators,

$$\frac{81}{197} > \frac{71}{197},$$

and comparing denominators,

$$\frac{71}{197} > \frac{71}{211}.$$

By the **transitivity** of order relations

$$\frac{81}{197} > \frac{71}{211}.$$

The strategy of constructing an intermediary that can be compared effortlessly to two objects is a recurrent theme in mathematics. See, for example, p. 184 in Volume 1 for the use of an intermediary in the context of length comparison.

Consider another example: which fraction is greater,

$$\frac{2}{7} \quad \text{or} \quad \frac{8}{12}?$$

This time, the greater-numerator smaller-denominator criterion does not apply, but comparison via an intermediary may still be more efficient than finding a common denominator.

Comparison by intermediary is not an algorithm—it is an unstructured exploratory method, much like the **chunking method** in division. For example, you could reformulate this problem in terms of cookies and children:

> Would you prefer sharing 2 cookies among 7 children, or sharing 8 cookies among 12 children?

It is unlikely that any cookie-loving child will go astray. It is clear, for example, that 7 children sharing 2 cookies with get less than half a cookie each, whereas 12 children sharing 8 cookies with get more than half a cookie each. To make this into a formal statement, we could write, for example,

$$\frac{2}{7} < \frac{2}{4} = \frac{1}{2} = \frac{8}{16} < \frac{8}{12}.$$

In words, 7 children sharing 2 cookies get fewer cookies each than 4 children sharing 2 cookies. 4 children sharing 2 cookies get the same amount of cookies each as 16 children sharing 8 cookies, which in turn get fewer cookies than 12 children sharing 8 cookies.

This example demonstrates a general pedagogical principle—the importance of being playful. Problems which seem to require technical procedures, at first sight, can often be solved with less effort by invoking intuitive metaphors. Finding a good metaphor, however, does not follow an algorithm; it requires creativity and imagination.

29.3 Fractions on the number line

The number line was introduced in Volume 1, p. 9, and used repeatedly for visualizing order relations and arithmetic operations. The inclusion of fractions into our number system calls for representing them on the number line as well.

The placement of numbers on the number line is dictated by the following two principles:

(a) **Order**: If a is less than b, then the point representing a is at the left of the point representing b.

(b) **Distance is proportional to difference**: Pairs of numbers whose differences are equal are placed at an equal distance apart.

The first stage in constructing a number line is to draw a line, place the number 0 at an arbitrary location on that line, and place the number 1 at an arbitrary point to its right:

Once the distance between 0 and 1 has been determined, the positions of 2, 3, etc. are imposed by the *distance proportional to difference* criterion:

Consider now the fraction ONE-HALF. Since

$$\frac{1}{2} + \frac{1}{2} = \frac{2}{2} = 1,$$

it follows that ONE-HALF is located half way between 0 and 1. Once we positioned ONE-HALF, its multiples

$$\frac{2}{2}, \quad \frac{3}{2}, \quad \frac{4}{2}, \quad \frac{5}{2}, \ldots$$

can be placed at fixed distances apart:

Similarly, since

$$\frac{1}{3} + \frac{1}{3} + \frac{1}{3} = \frac{3}{3} = 1,$$

it follows that ONE-THIRD is located at a point obtained by partitioning the segment between 0 and 1 into 3 equals parts. Once we positioned ONE-THIRD, its multiples,

$$\frac{2}{3}, \quad \frac{3}{3}, \quad \frac{4}{3}, \quad \frac{5}{3}, \ldots$$

can be placed at fixed distances apart:

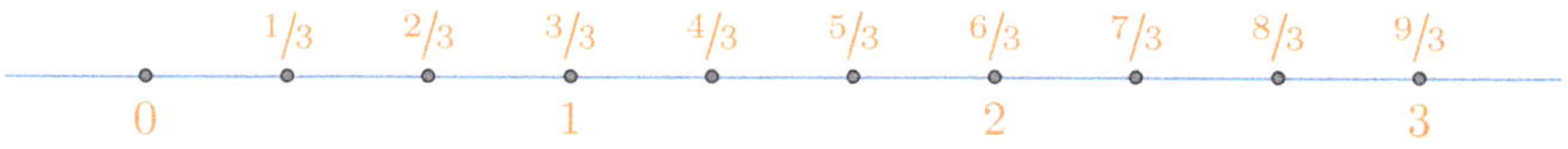

More generally, the position of the fraction $^m/_n$ is determined as follows: partition the segment between 0 and 1 into n equal parts; the length of each part represents a difference of $^1/_n$. Then, concatenate m such segments starting at 0 and proceeding to the right. The point thus reached represents the fraction $^m/_n$.

The following figure displays all the fractions with a denominator up to 8 on the unit segment. For readability, equivalent fractions (e.g., ONE-HALF and TWO-QUARTERS) are only displayed once:

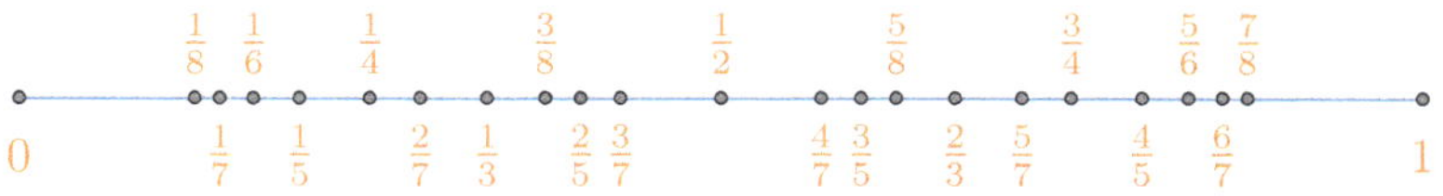

29.4 Density of the rational numbers

A defining property of the natural numbers is the existence of **successors** and **predecessors** (Volume 1, p. 8). The successor of a natural number is the smallest natural number among all those greater than it is. The predecessor of a natural number is the greatest natural number among all those less than it is. For example, 5 is the lowest natural number greater than 4 (hence its successor) and 8 is the greatest natural number less than 9 (hence its predecessor).

The existence of predecessors and successors does not carry through into the system of rational numbers. For example, there is no rational number that is the lowest among all the rational numbers greater than 4, nor there is a rational number that is the greatest among all the rational numbers less than TWO-THIRDS.

This property of the rational numbers can be formalized as follows: for every two rational numbers a and b satisfying $a < b$, there exists a third rational number c, satisfying

$$a < c < b.$$

This property may be more intuitive within the number line representation; it implies that every two points on the line have at least one other point in between.

The existence of an intermediate number can be applied recursively. It follows there also exists a rational number d, satisfying

$$a < d < c,$$

and a rational number e, satisfying

$$d < e < c,$$

and so on *ad infinitum*. In other words, there are infinitely many rational numbers between every two rational numbers. This property of the rational numbers is called **density**; we say that the rational numbers are dense.

29.5 McKay's theorem

The following anecdote occurred in an eighth grade class in the United States, and was publicized in 1973 by Laurence Sherzer (*The Mathematics Teacher* **66** pp. 229–230).

Sherzer's class was investigating various methods for finding a fraction c satisfying

$$a < c < b.$$

when given fractions a and b. As our running example, take the fractions ONE-SIXTH and ONE-FIFTH.

One of the methods presented in class was the following:

(a) Expand both fractions to get a common denominator:

$$\frac{1}{6} = \frac{5 \times 1}{5 \times 6} = \frac{5}{30} \qquad \text{and} \qquad \frac{1}{5} = \frac{6 \times 1}{6 \times 5} = \frac{6}{30}.$$

(b) If, in the expanded forms, the numerator of the greatest fraction is the successor of the numerator of the least fraction, expand both fractions by an additional factor of 2,

$$\frac{5}{30} = \frac{2 \times 5}{2 \times 30} = \frac{10}{60} \qquad \text{and} \qquad \frac{6}{30} = \frac{2 \times 6}{2 \times 30} = \frac{12}{60}.$$

(c) Construct a fraction using the denominator of the expanded fractions and the average of their numerators. The resulting fraction satisfies the required property,

$$\left(\frac{1}{6} =\right) \quad \frac{10}{60} < \frac{11}{60} < \frac{12}{60} \quad \left(= \frac{1}{5}\right).$$

While the class was practicing this method, one of the students, by the name of McKay, declared: "You dont have to work so hard. Just add numerator to numerator and denominator to denominator".

McKay's proposal was to take the fraction,

$$\frac{1+1}{5+6} = \frac{2}{11}$$

as an intermediary between ONE-SIXTH and ONE-FIFTH. Sherzer's first instinct was to reject the idea; adding numerator to numerator and denominator to denominator is a classical error in fraction addition. Nevertheless, he decided to open a debate whether this method was valid. First, the class verified that the relation

$$\frac{1}{6} < \frac{2}{11} < \frac{1}{5}$$

holds. Next, the class verified that McKay's proposal holds for other cases as well. To everyone's amazement, the method worked each time. At this stage, the teacher intervened and set to prove the correctness of McKay's proposal. He used algebraic means to prove that

$$\frac{a}{b} < \frac{c}{d},$$

implies

$$\frac{a}{b} < \frac{a+c}{b+d} < \frac{c}{d}.$$

McKay's proposal, known also as McKay's theorem, can be explained without resorting to algebraic means nor performing any calculation, and in a manner accessible to elementary school students. Consider the following setting:

> Teacher Malcolm distributed 2 cookies to 11 children. He shared one cookie equally among 5 children, and the other cookie sequally among the 6 remaining children.

The group of 6 had excellent reasons to feel deprived. They knew perfectly well that a whole divided by 6 is less than a whole divided by 5. So:

> The next day Teacher Malcolm distributed 2 cookies to 11 children evenly.

Clearly, the formerly privileged children got a smaller portion this time, whereas the formerly deprived children got a larger portion. That is,

$$\frac{2}{11} > \frac{1}{6} \qquad \text{and} \qquad \frac{2}{11} < \frac{1}{5},$$

which is exactly what we wanted to prove.

It should be noted that McKay's theorem was not a new discovery. The theorem is attributed to French mathematician Nicolas Chuquet (1455–1488). Sherzer's account is valuable not as a reference to a new theorem, but as a source for the potential surprises one may have in a classroom with unexpected student answers.

Mathematical problems

Problem 29.1. Place the number TWO-THIRDS on this number line:

Problem 29.2. Is it always true that a fraction whose numerator and denominator are prime numbers is irreducible?

Problem 29.3. Which fraction is larger,

$$\frac{12}{13} \quad \text{or} \quad \frac{13}{14}?$$

Justify your answer using an argument that requires no calculation.

Problem 29.4. Justify the inequalities

$$\frac{5}{8} < \frac{6}{9} < \frac{7}{10} < \frac{8}{11} < \cdots$$

using an intuitive argument.

Problem 29.5. Find three different numbers greater than TWO-SEVENTHS and less than ONE-THIRD.

Problem 29.6. Expand the following fractions,

$$\frac{1}{2}, \frac{1}{4}, \frac{1}{5}, \frac{1}{25}, \frac{1}{8}, \frac{1}{16}, \frac{1}{40} \quad \text{and} \quad \frac{1}{125},$$

such that the denominator is a decimal unit, i.e., either 1, 10, 100, 1000, etc. (This exercise is a good preparation for the study of **decimals**.)

Problem 29.7. Which fraction is greater,

$$\frac{3}{8} \quad \text{or} \quad \frac{4}{11}.$$

Solve using a metaphor of 8 children sharing 3 cookies versus 11 children sharing 4 cookies.

Pedagogical problems

Problem 29.8. Teacher Pazit asked her class to compare the fractions NINE-THIRTEENTHS and ELEVEN-FOURTEENTHS. Gili said that ELEVEN-FOURTEENTHS was larger, because $11 \times 13 = 132$ is greater than $9 \times 14 = 126$.

(a) Is Gili's answer correct?

(b) Does her method always work?
(c) Which of the comparison methods learned in this chapter does Gili's method relate to?

Problem 29.9. Teacher Merav asked her class to compare the improper fractions EIGHT-THIRDS and TEN-QUARTERS. Alon answered that both fractions are equal because both $8 \div 3$ and $10 \div 4$ yield a quotient of 2 with a remainder of 2. What would you reply?

Chapter 30

Fraction Addition and Subtraction

In the last three chapters we became acquainted with fractions, which are part of the extended number system of rational numbers. At this stage, we know what fractions are, how to represent them and how to use them to quantify parts of a whole. We are also proficient at comparing any two fractions, i.e., determining which is greater. This chapter, and the following two, are concerned with arithmetic operations involving rational numbers, in general, and fractions in particular. The present chapter focuses on the addition and subtraction of fractions. Chapter 31 studies fraction multiplication and Chapter 32 studies fraction division.

As advocated all along this book, the evaluation of arithmetic operations is only meaningful and valuable if one understands what is being evaluated. The meaning of all four arithmetic operations, each with its subtle variations, were explicated in Volume 1. Now that fractions enter into play, some of these meanings need revisiting. The main task of the next three chapters is to clarify the semantics of operations with fractions—to understand under what circumstances a certain operation has to be performed. Not surprisingly, the *how-to* will turn out to be much more meaningful, and therefore easier to grasp and remember, once the conceptual foundations have been laid.

Please note that despite my attempts to defer the study of arithmetic operations until a thorough understanding of the fraction concept has been attained, the addition of fractions has been around all along. General fractions, for example, were defined as the repeated addition of unit fractions. Similarly, mixed numbers, introduced on p. 191, were defined as sums of natural numbers and fractions. These operations were used without any explicit explanation, since they seem to be self-explanatory, at least in the special cases in which we used them.

Hurdles:

(a) Children do not relate the addition and subtraction of fractions to the meanings of these operations as joining together, taking apart, etc.

(b) Children add fractions by adding numerator to numerator and denominator to denominator.
(c) Children don't know how to validate the results of their evaluation.

Remedies:

(a) Practice addition and subtraction of fractions in the context of concrete word problems.
(b) Use visual aids to demonstrate addition and subtraction of fractions; vary the setting so that a fraction is sometimes interpreted as part of a whole and sometimes as part of a quantity.
(c) Address misconceptions by demonstrating them in concrete settings.
(d) Get children into the habit of estimating sums and differences of fractions prior to calculating them.

30.1 The meaning of fraction addition

Addition is an abstract arithmetic operation that can be viewed as a model for concrete operations on sets. Addition models the **union**, or the **joining together** of two sets; in less symmetric settings, it models **appending** one set to another (Volume 1, pp. 31–32). If a and b are natural numbers, then the value of the arithmetic expression

$$a + b$$

is the number of objects in a set formed by joining together a set of a objects and a set of b objects.

The meaning of addition as a model for joining together, or appending, does not change when extending the number system to rational numbers. The important thing to remember is that the operands may no longer quantify the number of whole objects in a set, but rather, might quantify parts of an object. As emphasized repeatedly in the previous chapters, an abstract expression acquires a concrete meaning once the abstract notion of a whole has been assigned a concrete sense.

Consider, for example, the three word problems:

(a) Mom gave Tal ONE-HALF of an apple. Dad gave Tal ONE-THIRD of an apple. What fraction of an apple did Tal end up with?

(b) ONE-HALF of the third grade students are at the gym and ONE-THIRD of them are in the auditorium. What fraction of the third grade students are either at the gym or in the auditorium?

(c) Tami had 50 dollars in her purse. She spent ONE-HALF of that sum buying a hat and ONE-THIRD of that sum buying a shirt. What fraction of her money did she spend?

In all three questions, the two given numbers quantify fractions of the same whole: an apple in the first example, the group of third graders in the second example, and the amount of money in Tami's purse in the third example. In all three questions, we are asked which fraction of the *same* whole is obtained when the two given fractions are joined together. Hence, the answer, in all three cases, is represented by the arithmetic expression

$$\frac{1}{2} + \frac{1}{3}.$$

Please note that the answer to Question (b) does not depend on the total number of children in third grade; the whole is the group of children, regardless of how many children it comprises. Likewise, the answer to Question (c) does not depend on the amount of money in Tami's purse; the whole is the amount of money, regardless of how much it is. That is, the data "50 dollars" is irrelevant to this question.

Even though the basic meaning of addition does not change with the incorporation of fractions, the interpretation of word problems requires special attention. The core of the difficulty, as is often the case with fractions, lies in possible ambiguities regarding the identity of the whole. As a demonstration, consider the word problem:

> ONE-HALF of the fourth grade students and TWO-THIRDS of the fifth grade students attended a party. What fraction of the students in grades 4 and 5 attended the party?

This problem features two numbers, ONE-HALF and TWO-THIRDS. The setting is that of joining together two groups and enquiring about the size of the compound group. This problem has all the features of addition, so it is tempting to express the solution by the arithmetic expression

$$\frac{1}{2} + \frac{2}{3}.$$

This is utterly wrong.

To pin down the source of the error, note that:

(a) The fraction ONE-HALF quantifies a part of the children in Grade 4.

(b) The fraction TWO-THIRDS quantifies a part of the children in Grade 5.

(c) The unknown answer quantifies a part of the total number of students in grades 4 and 5.

In other words, the addends and the sum refer to three different wholes, hence an addition sentence connecting these three numbers is devoid of any meaning.

Can the problem be solved at all? No. We must know how many children are in each grade in order to convert fractions of a class into an actual number of children. We will learn how to do so in Chapter 31.

When facing a problem that seems related to the addition of fractions, be sure that the following requirements are met:

(a) The two number which you are about to add are fractions of the same whole.
(b) The operation reflects either joining together or appending.
(c) The answer we are looking for is a fraction of the same whole as the two inputs.

If either of these conditions is not satisfied, you run the risk of obtaining an arithmetic expression that bears no relation to the problem at hand.

30.2 The meaning of fraction subtraction

Subtraction, the operation inverse to addition, assumes a number of different meanings. Subtraction is a model for **taking away**, **taking apart**, **comparison** and **complementation** (Volume 1, pp. 50–53).

These four meaning are exemplified in the following word problems:

(a) **Taking away**: Karen has 12 pencils, 3 of which have broken points. How many sharpened pencils does she have?
(b) **Taking apart**: Ayelet has 12 rabbits. How many male rabbits does she have if 3 of the rabbits are female?
(c) **Comparison**: There are 12 girls and 3 boys on Sharon's swim team. How many more girls than boys are there?
(d) **Complementation**: Dana has 3 dollars. How many more dollars does she need to purchase a 12-dollar shirt?

The solutions to all four problems are given by the difference

$$12 - 9,$$

whose value is the solution to the **unknown-addend equations**

$$3 + \boxed{?} = 12 \qquad \text{or} \qquad \boxed{?} + 3 = 12.$$

The same four interpretations are relevant when the operands are fractions. Consider the four word problems:

(a) **Taking away**: Karen had THREE-QUARTERS of a pound of flour. She made pancakes, using ONE-HALF of a pound. How many pounds of flour is she left with?
(b) **Taking apart**: Ayelet had THREE-QUARTERS of a pound of flour. She used ONE-HALF of a pound for making pancakes and the rest she used for making play dough. How many pounds of flour did she use for making play dough?
(c) **Comparison**: Sharon's pancake recipe requires THREE-QUARTERS of a pound of flour. Her friend's recipe requires ONE-HALF of a pound of flour. How many more pounds of flour does her recipe require?
(d) **Complementation**: Dana has ONE-HALF of a pound of flour. How many more pounds of flour does she need for making pancakes, if the recipe requires THREE-QUARTERS of a pound of flour?

The solutions to all four problems are given by the difference

$$\frac{3}{4} - \frac{1}{2},$$

whose value is the solution to the unknown-addend equations

$$\frac{1}{2} + \boxed{?} = \frac{3}{4} \qquad \text{or} \qquad \boxed{?} + \frac{1}{2} = \frac{3}{4}.$$

When faced with a problem that seems to relate to subtracting fractions, be sure that the following requirements are met:

(a) The two numbers which you are about to subtract are fractions of the same whole.
(b) The operation reflects either taking away, taking apart, comparison or complementation.
(c) The answer we are looking for is a fraction of the same whole as the the two inputs.

If either of these conditions is not satisfied, you run the risk of obtaining an arithmetic expression that bears no relation to the problem at hand.

To demonstrate this point, consider the word problem:

> THREE-QUARTERS of the sixth grade children came to a party. ONE-HALF of them left after an hour. What fraction of the sixth grade children remained after an hour.

On the surface, the setting is that of taking away. Yet, the fraction THREE-QUARTERS quantifies a part of *all* sixth grader, whereas the fraction ONE-HALF quantifies only a part of those who came to the party. Since the two fractions do not refer to the same whole, it makes no sense to solve this problem using plain subtraction; you have to wait until the next chapter to obtain the tools for solving this problem.

30.3 Evaluation of fraction addition and subtraction

Now that we know under what conditions addition and subtraction of fractions ought to be performed, we will learn how to evaluate the two operations. As we did in the previous chapter, we will open with evaluation methods based on concrete, hands-on models. Then, we will consider the particular case where both operands have like denominators. Finally, we will learn how to add and subtract fractions with unlike denominators.

30.3.1 Using concrete models

30.3.1.1 Addition

Consider the arithmetic expression:

$$\frac{1}{2} + \frac{1}{3},$$

whose value is the answer to the following question:

> What fraction of a whole is obtained when joining together ONE-HALF of that whole and ONE-THIRD of that whole?

A concrete model hinges on the choice of a whole. Following the tracks of ancient Greek mathematicians, let's take for whole a unit rod:

The addend ONE-HALF is modeled by dividing the unit rod into 2 equal segments, whereas the addend ONE-THIRD is modeled by dividing the unit rod into 3 equal segments:

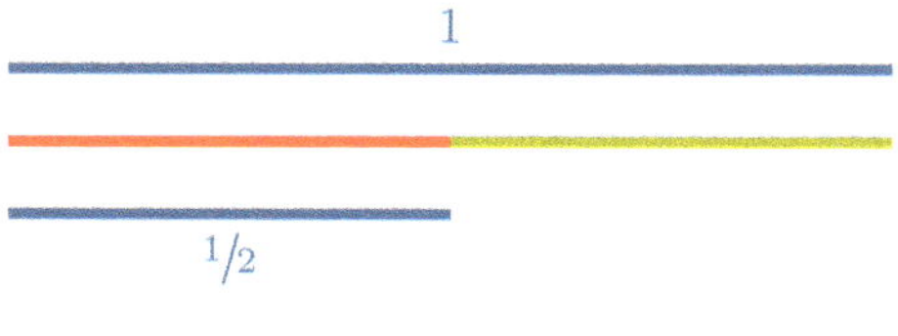

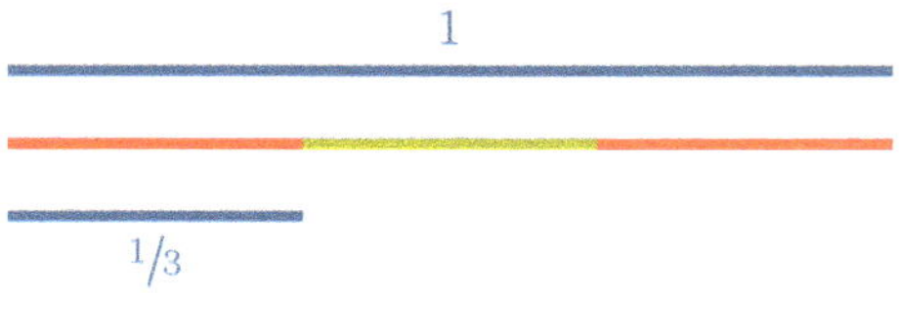

To represent the sum, we concatenate the two addends end-to-end, obtaining

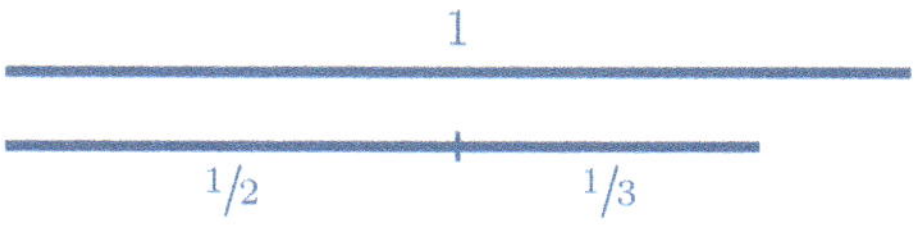

Alternatively, we may model the whole by a disc. The whole along with its two fractional parts are depicted below:

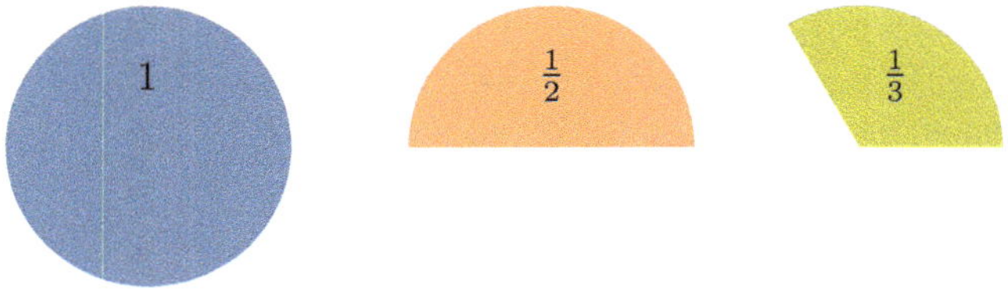

To represent the sum, we glue together the two sectors, obtaining

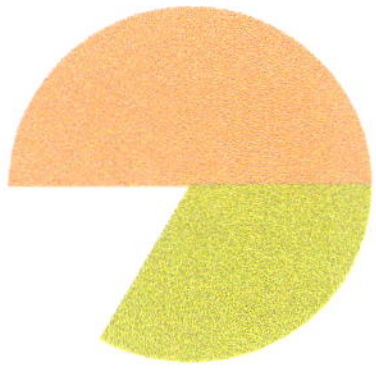

So what is the sum of ONE-HALF and ONE-THIRD? The two concrete models, the rod and the disc, reveal an intrinsic difficulty in modeling the addition of fractions: it is not easily discernible what fraction of the whole is represented by a given rod or by a given sector. It takes the following visualizations,

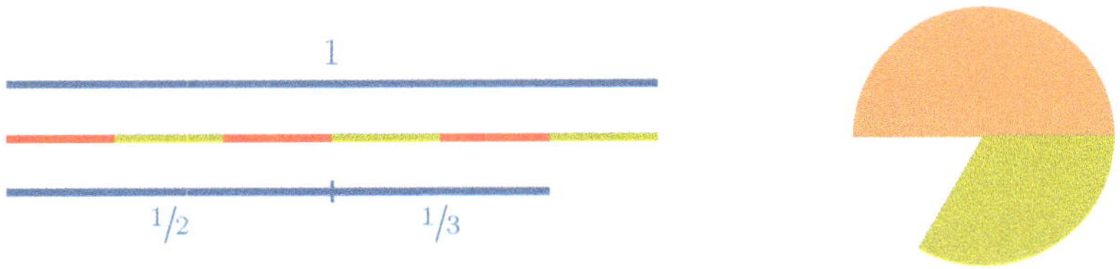

to identify that ONE-HALF of a whole plus ONE-THIRD of a whole make together FIVE-SIXTHS of a whole. This difficulty does not arise with natural numbers, where the only skill needed for carrying out addition in a concrete setting is counting—*if you can count, you can add.*

Nevertheless, I strongly encourage resorting to a diversity of concrete models, at all stages of learning fractions. Even in situations where the solution is not easily discernible, concrete models offer the following advantages:

(a) They sharpen one's conceptual perception of the operation.
(b) They help in making estimates. For example, the above models show convincingly that the sum of ONE-HALF and ONE-THIRD must be greater than THREE-QUARTERS and less than 1.

(c) They might suggest solution strategies, e.g., the search for a common denominator.

Another class of concrete models for the addition of fractions uses wholes that are sets of objects (see *fractions as part of a quantity* on p. 178). Suppose we rephrased the problem of adding ONE-HALF and ONE-THIRD as follows:

> What fraction of a group of children is obtained when joining together ONE-HALF of that group and ONE-THIRD of that group?

The key feature here is the use of objects that cannot be divided—children. For the question to make sense, one has to visualize a group of children of which both ONE-HALF and ONE-THIRD comprise a whole number of children, e.g., 6 children. Such a construction is at the core of evaluation methods using common denominators.

It is not hard to discover that a group of 6 children satisfies the requirements. ONE-HALF of this group consists of 3 children and ONE-THIRD of this group consists of 2 children. Both parts together comprise 5 children. Since each child constitutes ONE-SIXTH of the group, 5 children constitute FIVE-SIXTHS of the group, hence

$$\frac{1}{2} + \frac{1}{3} = \frac{5}{6}.$$

30.3.1.2 Subtraction

Consider the arithmetic expression:

$$\frac{1}{2} - \frac{1}{3},$$

whose value can be viewed as the answer to the following question:

> What fraction of a whole should be added to ONE-THIRD of that whole to obtain ONE-HALF of that whole?

Note that here we chose to interpret subtraction as complementation.

Using our rod model, we can visualize the difference as follows:

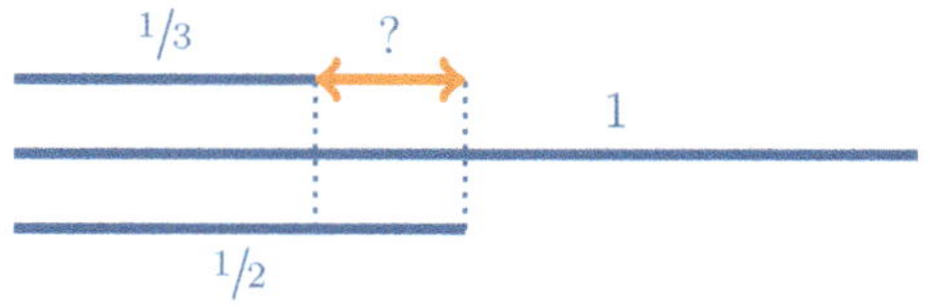

Here too, the answer is not easily discernible, unless presented with the following visual aid:

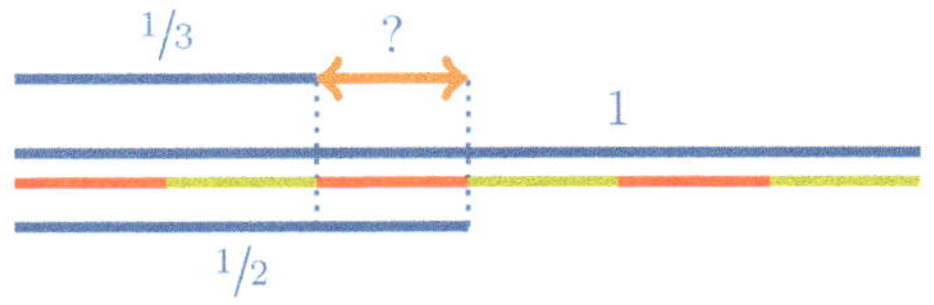

which makes evident the fact that the difference between ONE-HALF of a whole and ONE-THIRD of a whole is ONE-SIXTH of a whole.

30.3.2 Fractions with like denominators

Just as with fraction comparison, the evaluation of fraction addition and fraction subtraction is straightforward when both operands have the same denominator.

30.3.2.1 Addition

Consider the word problem:

> The Peled Family ordered a pizza. Roee ate THREE-EIGHTHS of the pizza and Iddo ate TWO-EIGHTHS of the pizza. What fraction of the pizza was eaten?

Before rushing to write a number sentence, let's ascertain that all prerequisites for fraction addition are satisfied:

(a) Both fractions, THREE-EIGHTHS and TWO-EIGHTHS, quantify parts of the *same whole*—a pizza.
(b) The answer we are looking for is the outcome of *joining together* the two parts.
(c) The answer we are looking for quantifies a part of the *same whole*.

Thus, the solution is the value the arithmetic expression

$$\frac{3}{8} + \frac{2}{8}.$$

The value of this sum is readily obtained if we recall that general fractions are repeated addition of unit fractions with the same denominator. Thus,

$$\frac{3}{8} + \frac{2}{8} = \left(\frac{1}{8} + \frac{1}{8} + \frac{1}{8}\right) + \left(\frac{1}{8} + \frac{1}{8}\right).$$

Since addition is **associative**, we may remove the parentheses, yielding by definition,

$$\frac{3}{8} + \frac{2}{8} = \frac{1}{8} + \frac{1}{8} + \frac{1}{8} + \frac{1}{8} + \frac{1}{8} = \frac{5}{8}.$$

Alternatively, we may invoke the **distributive property**, and write the following identities,

$$3 \times \frac{1}{8} + 2 \times \frac{1}{8} = (3+2) \times \frac{1}{8} = 5 \times \frac{1}{8} = \frac{5}{8}.$$

Please note that even though we are adding fractions, the use of the distributive property turns the problem into that of evaluating a sum of natural numbers, 3 plus 2. Indeed, suppose we had reformulated the problem as follows:

> The Peled Family ordered a pizza. Roee ate 3 eighths-of-a-pizza and Iddo ate 2 eighths-of-a-pizza . How many eighths-of-a-pizza did both eat together?

This subtle reformulation, in which rather than quantifying parts of a pizza (i.e., the whole is a pizza) we quantify eighths-of-a-pizza, turns the crux of the problem into one of adding natural numbers. The transition from *whole=pizza* into *whole=eighth-of-a-pizza* was made possible because both fractions have the same denominator, 8.

The above considerations are not restricted to the particular problem in hand. Let k, m and n be natural numbers. Since

$$\underbrace{\left(\frac{1}{n} + \frac{1}{n} + \cdots + \frac{1}{n}\right)}_{k \text{ times}} + \underbrace{\left(\frac{1}{n} + \frac{1}{n} + \cdots + \frac{1}{n}\right)}_{m \text{ times}} = \underbrace{\frac{1}{n} + \frac{1}{n} + \cdots + \frac{1}{n}}_{k+m \text{ times}},$$

it follows that when adding fractions with like denominators, the sum is a fraction having the same denominator and a numerator equal to the sum of the two numerators:

> For every three natural numbers k, m and n
>
> $$\frac{k}{n} + \frac{m}{n} = \frac{k+m}{n}.$$

30.3.2.2 Subtraction

Consider the word problem:

> FIVE-NINTHS of the country's population live in cities whereas TWO-NINTHS of the country's population live in farming communities. How much larger is the fraction of the urban population than the fraction of rural population?

As we did in the previous section, let's first ascertain that all prerequisites for the subtraction of fractions are satisfied:

(a) Both fractions, FIVE-NINTHS and TWO-NINTHS quantify parts of the *same whole*—the country's population.
(b) The answer we are looking for is the outcome of *comparing* both parts.
(c) The answer we are looking for quantifies a part of the *same whole.*

Thus, the answer is the value the arithmetic expression

$$\frac{5}{9} - \frac{2}{9}.$$

We can take advantage of both fractions having the same denominator, and turn the problem into evaluating a difference of whole numbers:

> 5 ninths-of-the-country's-population live in cities whereas 2 ninths-of-the-country's-population live in farming communities. How many more ninths-of-the-country's-population are living in cities compared to that living on farms?

The solution is readily found: there are 3 ninths-of-the-country's-population more living in cities, i.e.,

$$\frac{5}{9} - \frac{2}{9} = \frac{3}{9}.$$

Without further ado, we can formulate the general rule: Let k, m and n be natural numbers, with $k > m$. Since

$$\underbrace{\left(\frac{1}{n} + \frac{1}{n} + \cdots + \frac{1}{n}\right)}_{k \text{ times}} - \underbrace{\left(\frac{1}{n} + \frac{1}{n} + \cdots + \frac{1}{n}\right)}_{m \text{ times}} = \underbrace{\frac{1}{n} + \frac{1}{n} + \cdots + \frac{1}{n}}_{k-m \text{ times}},$$

it follows that when subtracting fractions with like denominators, the difference is a fraction having the same denominator and a numerator equal to the difference between the two numerators:

> For every three natural numbers k, m and n
> $$\frac{k}{n} - \frac{m}{n} = \frac{k-m}{n}.$$

30.3.3 Fractions with unlike denominators

30.3.3.1 Addition

Consider once again the arithmetic expression

$$\frac{1}{2} + \frac{1}{3}.$$

In Section 30.3.1.1, we visualized the sum using two concrete models, and saw that the value of that sum is not easily discernible, unless a visual hint is provided. The core of this difficultly lies in both fractions having unlike denominators: we are asked to join together a fraction of a whole obtained by partitioning the whole into 2 equal parts with a fraction of a whole obtained by partitioning the whole into 3 equal parts. It is not a priori clear what partition of a whole is needed to generate the sum.

The addition of fractions with unlike denominators turns out to be a stumbling block for many children. The most common error is adding numerator to numerator and denominator to denominator, resulting in

$$\frac{1+1}{2+3} = \frac{2}{5}.$$

An examination of the concrete models on p. 221 would reveal right away the fault of this answer.

To correctly add fractions with unlike denominators, we assemble two skills which were already acquired:

(a) Converting every two fractions into equivalent forms having a common denominators (p. 204).
(b) Adding fractions with like denominators (p. 223).

The fractions ONE-HALF and ONE-THIRD can be expanded as follows:

$$\frac{1}{2} = \frac{3 \times 1}{3 \times 2} = \frac{3}{6} \qquad \text{and} \qquad \frac{1}{3} = \frac{2 \times 1}{2 \times 6} = \frac{2}{6}.$$

Hence

$$\frac{1}{2} + \frac{1}{3} = \frac{3}{6} + \frac{2}{6} = \frac{5}{6}.$$

This solution strategy is always applicable. Note however how many steps it entails: (i) find a common multiple of the two denominators; (ii) find the corresponding expansion factor of each fraction; (iii) perform two fractions expansions; (iv) add two fractions with like denominators. Moreover, it is often beneficial to simplify the final solution by converting it into an irreducible form. Learning to implement each of those steps takes time and practice, and is best internalized when accompanied by concrete models.

30.3.3.2 Subtraction

Evaluating the difference of fractions with unlike denominators follows the same steps as evaluating their sum:

(a) Bring the two fractions into equivalent forms having a common denominator.
(b) Subtract the resulting fractions, which now have like denominators.

Let's exemplify this solution strategy on the evaluation of the difference

$$\frac{3}{4} - \frac{2}{5}.$$

A common multiple (in fact, the least common multiple) of 4 and 5 is 20, hence the two operands can be expanded as follows:

$$\frac{3}{4} = \frac{5 \times 3}{5 \times 4} = \frac{15}{20} \qquad \text{and} \qquad \frac{2}{5} = \frac{4 \times 2}{4 \times 5} = \frac{8}{20}.$$

Hence,

$$\frac{3}{4} - \frac{2}{5} = \frac{15}{20} - \frac{8}{20} = \frac{7}{20}.$$

30.4 Fractions in ancient Egypt

A valuable source regarding the history of mathematics was revealed in a papyrus dating from 1650 BC, and purchased in 1858 by a Scottish antiquities dealer, Alexander Rhind. This papyrus is presently on display in London's British Museum.

The Rhind papyrus revealed to modern civilization how ancient Egyptians used to denote fractions. The Egyptians only had signs for unit fractions, that is, only for those fractions representing the size of a share when partitioning a whole into an integer number of equally-sized parts. The Egyptian's numerals for unit fractions are akin to our contemporary notation. A unit fraction was denoted by a two-layered numeral: At the top of the numeral stood a sign representing the whole, possible a sign of a mouth,

The Egyptian mouth is analogous to the numerator 1 in contemporary notation. At the bottom of the numeral were hieroglyphs representing a whole number—our contemporary denominator. For example,

denoted, from left to right, the unit fractions ONE-THIRD and ONE-TENTH (see Volume 1, p. 22 for the hieroglyphic representation of natural numbers).

Fraction were represented as sums of unit fractions. Technically speaking, this is still the case today, except that we have a short-hand notation for the repeated addition of unit fractions. Egyptian notation, however, required all the fraction composing the sum to be distinct. For example, the fraction TWO-THIRDS could not be represented as ONE-THIRD plus ONE-THIRD; a legitimate notation was

$$\frac{1}{2}+\frac{1}{6}.$$

The Egyptian notation for fractions raises two immediate questions:

(a) Do fractions have a unique representation as a sum of non-repeating unit fractions?
(b) Does every fraction have a representation in the Egyptian system?

The answer to the first question is negative. Take the fraction TWO-THIRDS, for example. In addition to ONE-HALF plus ONE-SIXTH, the following are two other legitimate representations,

$$\frac{1}{3}+\frac{1}{4}+\frac{1}{12} \qquad \text{and} \qquad \frac{1}{2}+\frac{1}{10}+\frac{1}{15}.$$

The answer to the second question is positive—the Egyptians had a system of notation capable of expressing every fraction. Nevertheless, it is not known even today how many non-repeating unit fractions are needed in order to represent a general fraction.

Mathematical problems

Problem 30.1. On a basketball team, Player A scores ONE-HALF of his free shots whereas Player B scores ONE-THIRD of his free shots. Coach Orly wants to know what fraction of their shots both players score together.

(a) Is this a problem of fraction addition?

(b) Can you provide Coach Orly with an answer?

Problem 30.2. Evaluate the following differences:

$$\frac{1}{2}-\frac{1}{3}$$
$$\frac{1}{3}-\frac{1}{4}$$
$$\frac{1}{4}-\frac{1}{5}$$
$$\frac{1}{5}-\frac{1}{6}$$

Can you detect a pattern? Can you explain this pattern using a concrete model?

Problem 30.3. Find three natural numbers, a, b and c satisfying

$$\frac{1}{a}+\frac{1}{b}+\frac{1}{c}=1.$$

Find three different solutions; interchanging variables does not count as an additional solution. Explain why there are no more solutions.

Problem 30.4. How would the fraction FIVE-SEVENTHS be represented in ancient Egypt?

Problem 30.5. Place the numbers 3, 4, 5 and 6 in the following template

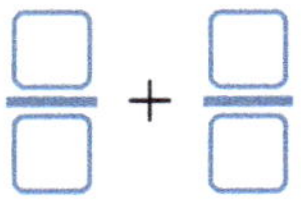

to obtain (a) a maximal sum (b) a minimal sum.

Pedagogical problems

Problem 30.6. Teacher Cam asked her class: "How much is the sum of ONE-HALF of an hour and ONE-THIRD of an hour?". Roni replied that the answer was 50 minutes, but encountered a difficulty when trying to write an appropriate number sentence.

(a) Is Roni's answer correct?
(b) Why wasn't Roni able to write an appropriate number sentence?
(c) How would you help Roni resolve this problem?

Chapter 31

Fraction Multiplication

This chapter presents multiplication of rational numbers—natural numbers and fractions. Fraction multiplication is a deceivingly simple operation. At the stage where it is taught—in fifth and sixth grades—the evaluation of products of fractions is a relatively easy task; one has to multiply numerator by numerator and denominator by denominator to obtain the desired product. However, as emphasized repeatedly throughout this book, knowing *how* to evaluate an arithmetic operation is far less important than knowing *when* to evaluate it, and *why* it is evaluated in this way.

Products of natural numbers can always be interpreted as repeated addition. Products of fractions, however, require a different interpretation. Thus, the multiplication concept must be extended and generalized in order to make sense of fraction multiplication. The main focus of this chapter is acquiring a unified image of multiplication, applicable to both natural numbers and fractions, and extending the scope of problems modeled by multiplication. Evaluation strategies are then derived based on our ability to correctly interpret products of fractions.

Hurdles:

(a) Children don't understand the meaning of fraction multiplication and are content with knowing how to evaluate products.

(b) Children are confused by the verbal interpretation of the multiplication sign as *times* in the context of whole numbers and *of* in the context of fractions.

(c) Children have a misconception whereby multiplication is always a magnifying operation.

Remedies:

(a) Explain the meaning of multiplication as a scaling operation; this interpretation is sensible for both whole numbers and for fractions.
(b) Practice fraction multiplication in the context of concrete word problems.
(c) Use visual aids to demonstrate fraction multiplication.
(d) Emphasize the relation between products of whole numbers, products of fractions and whole numbers and products of fractions.
(e) Use concrete models to demonstrate why multiplication by a fraction is not magnifying, but rather reducing.

31.1 Numbers and multiplication revisited

The number concept was first introduced in this book in Chapter 1, and was further elaborated in subsequent chapters. Notably, fractions were introduced in Chapter 27, extending the number system, which at first consisted only of natural numbers.

Multiplication was first introduced in Chapter 7, building upon the notion of **repeated addition**. We encountered various settings in which multiplication models concrete problems, such as counting combinations and measuring area. In the context of fractions, products of natural numbers and a unit fractions were used to define general fractions. That is, the fraction $^m/_n$ is defined as a repeated addition, m times, of the unit fraction $^1/_n$,

$$\frac{m}{n} = m \times \frac{1}{n}.$$

In this section we revisit both concepts of numbers and multiplication, showing that numbers and multiplication are intimately related concepts. Understanding this relation will help us build a coherent image of fraction multiplication.

31.1.1 The number concept revisited

Numbers are abstract entities. They acquire a concrete meaning only once a *whole*, to which we attribute the value ONE, has been specified. For example, the natural number 5 is an abstract entity, but 5 elephants are concrete, and so are 5 fingers, 5 chicks and 5 continents. The common property of all those 5 "somethings" is that they are sets of objects that can be matched to each other in a one-to-one manner (Volume 1, p. 2).

The concretization of fractions, just like the concretization of natural numbers, relies on the specification of a whole. The number ONE-HALF is an abstract entity, but ONE-HALF of a peach is concrete, and so are ONE-HALF of a pound of flour, ONE-HALF of an hour and ONE-HALF of a group of children. The common property of all those ONE-HALF of "something" is that if we join 2 such halves together, we obtain concrete objects, each of which designated as a *whole*. It is of primary importance to remember that the designation of a certain object as a whole—whether an elephant, a finger, a chick, a continent, a peach, a pound of flour, an hour or a group of children—is a matter of choice, and varies from one context to another. For example, a pound of flour can be a whole in one setting, and it can be 2 wholes or ONE-HALF of a whole in another setting—it all depends on which problem we are trying to solve.

We have already observed that natural numbers can be visualized as replicating machines, or in mathematical jargon, **operators**. As explained in Chapter 3, an operator can be visualized as a machine taking (possibly multiple) inputs and retuning a single output. Natural number, viewed as operators take a single input, which may be an object, part of an object or multiple objects, and return an output, which may as well be an object, part of an object or multiple objects.

Let's consider examples. The number 4 can be visualized as an operator, which we will call FOUR OF. Whatever its input is, its output is 4 copies of that input.

If the input is an apple, the output is 4 copies of that apple:

And if the input is a group of 2 children, the output is 4 copies of a group of two children:

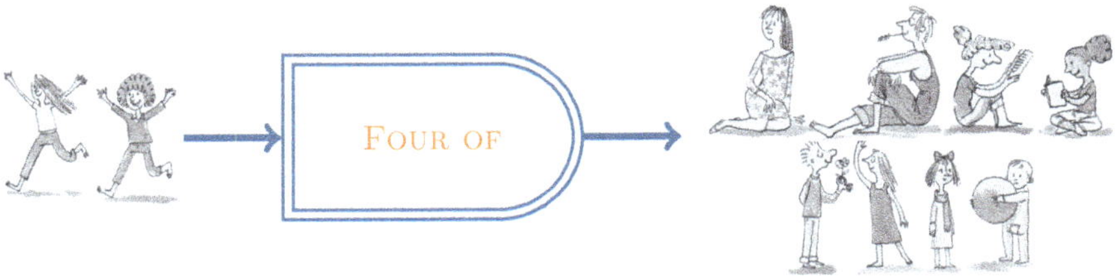

Similarly, the number ONE-HALF can be visualized as an operator, which we will call ONE-HALF OF. Whatever its input is, its output is a fraction of its input obtained by dividing it into 2 equal parts.

If the input is an apple, the output is a fraction of that apple obtained by dividing it into 2 equals parts:

And if the input is a group of 2 children, the output is fraction of that group obtained by dividing it into 2 equals parts:

We will term operators such as FOUR OF and ONE-HALF OF **numerical operators**. Every number, whether a natural number or a fraction, can be construed as a numerical operator. A numerical operator corresponding to a natural number n returns n copies of its input, whereas a numerical operator corresponding to a fraction m/n returns a fraction m/n of its input.

You may think that the notion of numerical operators is merely a technical exercise. However, children's first acquaintance with numbers is usually as numerical operators. At first, numbers are always attached to nouns, acting as replicating machines. Children's understanding of the abstract number FOUR is a consequence of a prolonged experience with concrete notions of FOUR objects. Pre-schoolers would have a hard time reasoning about the number FOUR without asking "FOUR what?". This suggests that the mental image associated with numbers at early age is similar to that of numerical operators. FOUR turns any object into 4 copies of that object.

The same mental process repeats itself several years later at the first encounter with fractions. Children who are already able to reason abstractly about natural numbers find it difficult to reason abstractly about fractions: there is no ONE-HALF without asking "ONE-HALF of what?". For children who have just started learning fractions, ONE-HALF is still a numerical operator whose output is ONE-HALF of its input. It takes several years of concrete experience to create in one's mind an abstract notion of a fraction.

31.1.2 Multiplication revisited

After presenting an alternative image of numbers—numerical operators—we revisit the multiplication operation. For concreteness, we will examine four word problems, each corresponding to a different choice of numerical inputs, alternating between natural numbers and fractions.

Problem 1: Product of natural numbers

A carton contains 12 eggs. How many eggs are in 3 cartons?

This is a typical multiplicative scene. The solution to Problem 1 is represented by the product

$$3 \times 12.$$

We could rephrase the solution as follows:

3 cartons of eggs contain 3 times 12 times an egg.

Using our construct of numerical operators, the content of 3 cartons of eggs is the output of two serially-connected numerical operators: the numerical operator TWELVE OF takes an egg and replicates it 12 times; the numerical operator THREE OF takes the output of the first operator and replicates it 3 times:

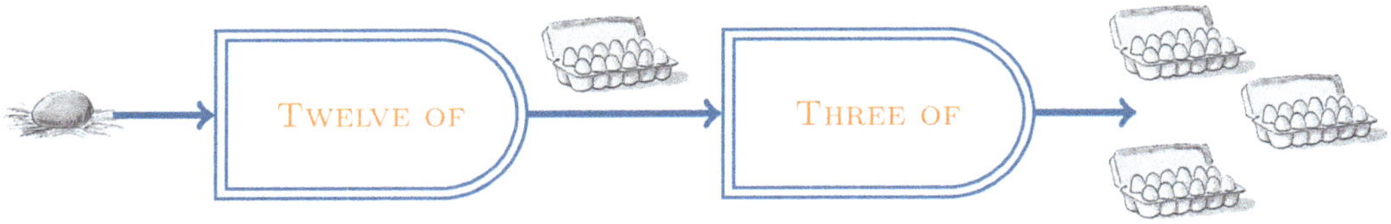

The multiplication of two numbers is embodied in a serial network of two numerical operators. The multiplier is "downstream" of the multiplicand, implying that the numerical operator corresponding to the multiplicand acts first.

Viewed under this angle, what do we need to do to solve Problem 1? How does the evaluation of a product relate to numerical operators?

The answer is depicted below:

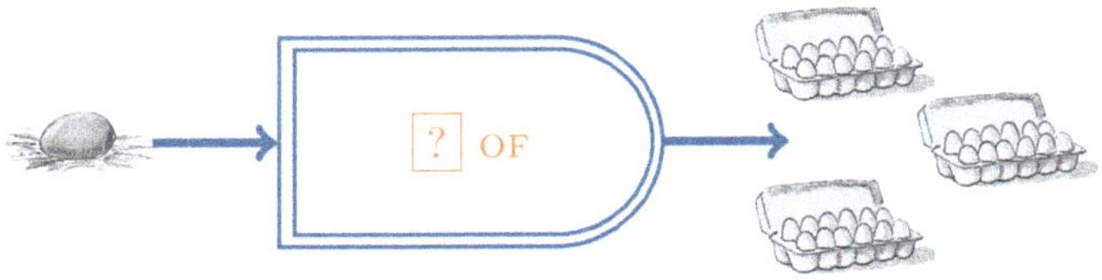

When we evaluate a product, we find *one* numerical operator that is equivalent to *two* serially-connected operators. The number sentence

$$3 \times 12 = 36$$

is illustrated in the form of a "numerical operator sentence":

Problem 2: Product of a fraction by a natural number

> TWO-SEVENTHS of a pound of flour are required to prepare a quiche. How many pounds of flour are required to prepare 3 quiches?

Adopting the viewpoint of numerical operators, the numerical operator TWO-SEVENTHS OF takes for input a pound of flour and returns the amount of flour needed for making one quiche. The input for the numerical operator THREE OF is the amount of flour needed for making one quiche; the output is the amount of flour needed for making 3 quiches.

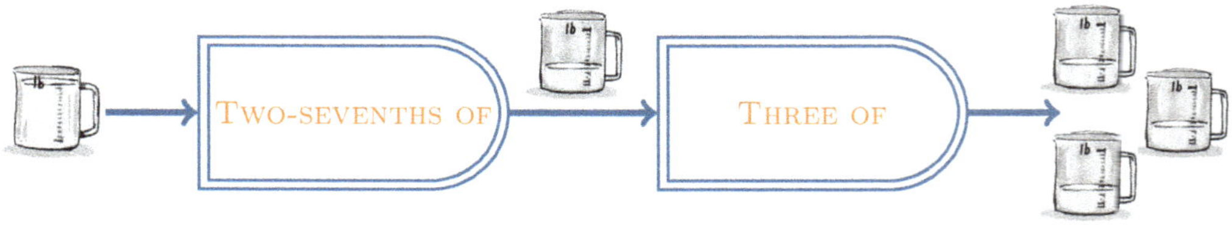

We seek a numerical operator equivalent to this serial network of numerical operators:

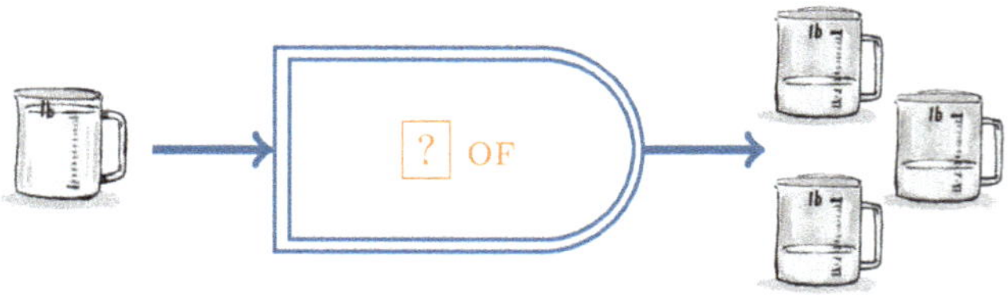

Problem 1 and Problem 2 have the exact same structure; the difference is that in the latter the multiplicand is a fraction, rather than a natural number. The first numerical operator returns part of its input rather than multiple copies of its input.

Since, Problem 2 has the same structure as Problem 1—we seek a numerical operator equivalent to a serial network of numerical operators—the arithmetic operation modeling Problem 2 is the same as the arithmetic operation modeling Problem 1; in both cases, the solution is represented by a product. The solution to Problem 2 is represented by the product

$$3 \times \frac{2}{7}.$$

Problem 3: Product of a natural number by a fraction

> There are 30 children in a class. TWO-FIFTHS of them are in the backyard. How many children are in the backyard?

Adopting the metaphor of numerical operators, the operator THIRTY OF takes for input a child and replicates it 30 times, yielding a group of children which we call a class. The input for the numerical operator TWO-FIFTHS OF is the output of the THIRTY OF operator; it returns the group of children that are in the backyard:

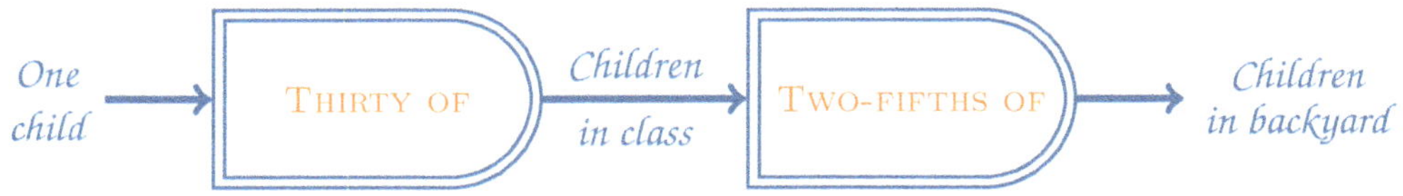

We seek a numerical operator equivalent to this serial network of numerical operators:

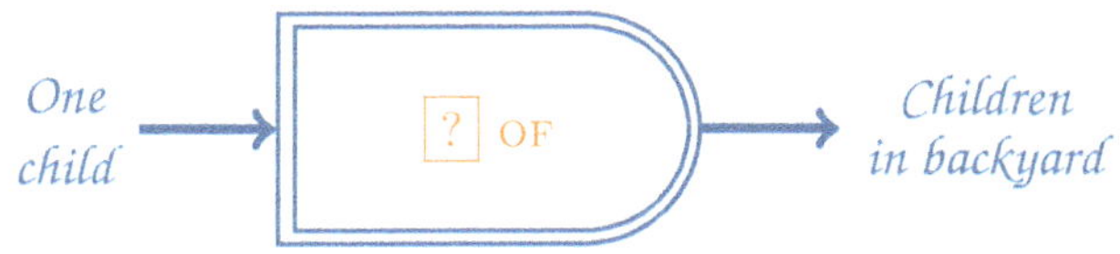

In Problem 3, the multiplicand is a natural number, 30, and the corresponding numerical operator returns multiple copies of its input. The multiplier is a fraction, TWO-FIFTHS, hence the corresponding numerical operator returns a part of its input.

Problem 3 has the same structure as Problems 1 and 2; the difference is that in the latter the multiplier is a fraction and the multiplicand is a natural number. Therefore, the same arithmetic operation models all three problems, and the solution to Problem 3 is represented by the product

$$\frac{2}{5} \times 30.$$

Problem 4: Product of fractions

> THREE-QUARTERS of the campers were assigned to help in decorating camp. Out of those, TWO-THIRDS were asked to hang posters. What fraction of the campers ended up hanging posters?

Adopting the metaphor of numerical operators, the operator THREE-QUARTERS OF takes for input the group of campers and returns that part of the group assigned to decorate. The numerical operator TWO-THIRDS OF takes for input the output of the THREE-QUARTERS OF operator, and return the group of campers that are hanging posters:

We seek a numerical operator equivalent to this serial network of numerical operators:

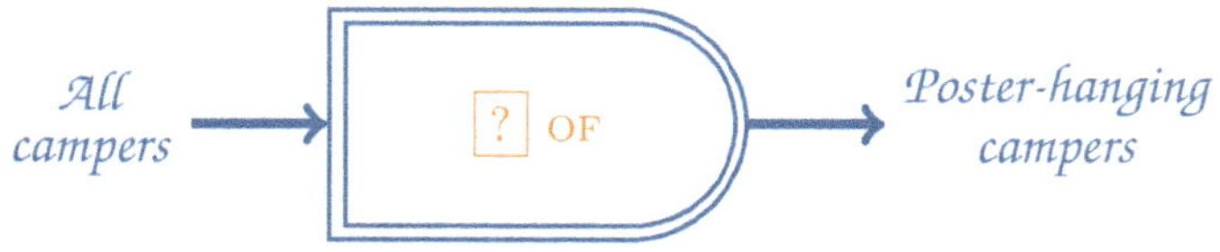

In Problem 4, both the multiplicand and the multiplier are fractions, hence both numerical operators return a part of their input.

Problem 4 has the exact same structure as Problems 1 through 3—we look for a numerical operator equivalent to a serial network of numerical operators. Since all four problems have the same structure, the same arithmetic operation models the four solutions. The solution to Problem 4 is represented by the product

$$\frac{2}{3} \times \frac{3}{4}.$$

Recapitulation

Problems 1 through 4 present four different word problems having the exact same structure; they only differ in whether the operands are natural numbers or fractions:

Problem 1: The product is a plurality of a plurality of the whole—3 times 12 eggs.

Problem 2: The product is a plurality of a fraction of the whole—3 times TWO-SEVENTHS of a pound of flour.

Problem 3: The product is a fraction of a plurality of the whole—TWO-FIFTHS of 30 children.

Problem 4: The product is a fraction of a fraction of the whole—TWO-THIRDS of THREE-QUARTERS of the campers.

Since the four problems have the same structure, they are all modeled by the same arithmetic operation. In all cases the solution is represented by the product of the two operands.

31.2 Evaluating products

Now that we are acquainted with a selection of settings modeled by the product of two rational numbers, we turn to consider the evaluation of such products. Since products of natural numbers have already been studied in depth in Chapters 7 and 19, we will focus on products in which at least one of the two operands is a fraction. To operate with concrete examples in mind, we will solve Problems 2 through 4 presented in the previous section.

31.2.1 Fractions multiplied by natural numbers

We start by examining products in which the multiplicand is a fraction and the multiplier is a natural number. Recall Problem 2:

> TWO-SEVENTHS of a pound of flour are required to prepare a quiche. How many pounds of flour are required to prepare 3 quiches?

The solution is represented by the product

$$3 \times \frac{2}{7}.$$

In this problem, the multiplicand is TWO-SEVENTHS and the multiplier is 3. 3 times a whole equals that whole added 3 times. In this word problem, this whole to which the multiplier 3 refers is TWO-SEVENTHS of some other whole—a pound a flour; recall how we have stressed repeatedly that *whole* may refer to anything, including a fraction of another whole!

Thus, multiplication can be interpreted as repeated addition:

$$3 \times \frac{2}{7} = \frac{2}{7} + \frac{2}{7} + \frac{2}{7},$$

displaying a sum of three addends, all fractions with a common denominator. Applying the addition rule for fractions with like denominators (p. 223),

$$3 \times \frac{2}{7} = \frac{2+2+2}{7} = \frac{3 \times 2}{7} = \frac{6}{7}.$$

The solution to Problem 2 is:

> It takes SIX-SEVENTHS of a pound of flour to prepare 3 quiches.

This method for multiplying a fraction by a natural number is not restricted to the numbers 3 and TWO-SEVENTHS. Let k, m and n be any natural numbers. Using the interpretation of multiplication as repeated addition, and applying the addition rule for fractions with like denominators:

$$k \times \frac{m}{n} = \underbrace{\frac{m}{n} + \frac{m}{n} + \cdots + \frac{m}{n}}_{k \text{ times}} = \frac{\overbrace{m + m + \cdots + m}^{k \text{ times}}}{n},$$

from which we obtain the general formula:

For every three natural numbers k, m and n:

$$k \times \frac{m}{n} = \frac{k \times m}{n}.$$

31.2.2 Natural numbers multiplied by fractions

We proceed to consider products in which the multiplicand is a natural number and the multiplier is a fraction. This setting is experienced as harder than the previous setting, because one can no longer interpret multiplication as repeated addition. To advance gradually, we first examine the particular case in which the multiplier is a unit fraction.

31.2.2.1 Natural numbers multiplied by unit fractions

Consider a variant of Problem 3, which we will call Problem 3a:

> There are 30 children in a class. ONE-FIFTH of them are in the backyard. How many children are in the backyard?

The solution is represented by the product

$$\frac{1}{5} \times 30.$$

We have to determine how many children are ONE-FIFTH of 30 children.

ONE-FIFTH of a whole—any whole—is a part of the whole obtained when dividing this whole into 5 equal shares. In the present case, the whole is 30 children, hence ONE-FIFTH of 30 children is a part of that group of children obtained when dividing them into 5 equal groups. The outcome of 30 objects (children) partitioned into 5 sets is represented by the quotient $30 \div 5$. Hence,

$$\frac{1}{5} \times 30 = 30 \div 5.$$

The quotient on the right is readily evaluated, 6, hence the solution to Problem 3a is:

> There are 6 children in the backyard.

Here too, we used considerations that are not restricted to the particular numbers in hand. Hence, we can formulate a general rule: for every pair of natural numbers k and n,

$$\frac{1}{n} \times k = k \div n.$$

Moreover, as we learned in Chapter 28, the quotient on the right-hand side equals the fraction $^k/_n$, hence

For every two natural numbers k and n:

$$\frac{1}{n} \times k = \frac{k}{n}.$$

Comments

(a) By evoking the definition of a fraction as quantifying part of a whole, we showed that multiplication by ONE-FIFTH is the same as division by 5. More generally, multiplication by $1/n$ is the same as division by n. Children know from a young age that ONE-HALF OF and DIVIDED BY TWO are synonymous. Their difficulty lies in relating this fact to a number sentence. Many children have a hard time identifying ONE-HALF OF with ONE-HALF TIMES.

(b) On the go, we proved that the product of a natural number and a fraction satisfies the **commutative property**. Indeed, in the previous section we showed that

$$k \times \frac{1}{n} = \frac{k}{n},$$

whereas in this section we showed that

$$\frac{1}{n} \times k = \frac{k}{n}.$$

One could argue that this is not something worth mentioning, as we already know that multiplication is commutative. This argument should be taken with a grain of salt. In Chapter 7, p. 82, we *proved* that multiplication is commutative using a rectangular array model, which relied explicitly on the meaning of multiplication as repeated addition. Why should commutativity hold for fraction-valued factors? Is it obvious that ONE-FIFTH of 30 is the same as 30 times ONE-FIFTH?

The answer is illustrated below:

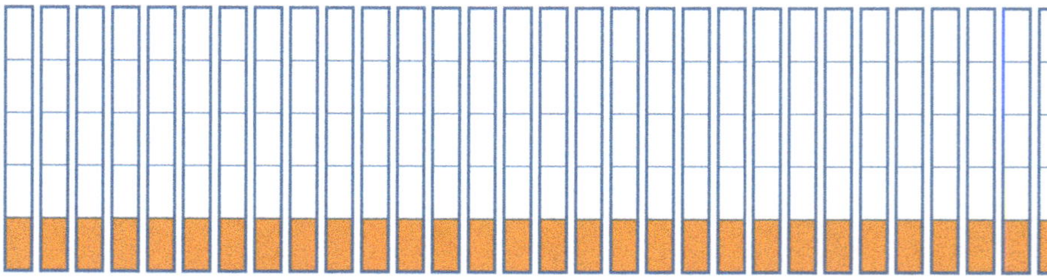

This figure displays 30 rectangles, each of which is partly colored orange. On the one hand, the total colored part is 30 times ONE-FIFTH of a rectangle. On the other hand, it constitutes ONE-FIFTH of the whole figure, i.e., ONE-FIFTH of the 30 rectangles, is colored. Hence,

$$\frac{1}{5} \times 30 = 30 \times \frac{1}{5}.$$

(c) The fact that

$$\frac{1}{5} \times 30 = 30 \div 5$$

is consistent with the **distributive property**. Let's denote the unknown value of $1/5 \times 30$ by the symbol x (the standard symbol for an unknown). Add x to itself 5 times; by the distributive property,

$$x + x + x + x + x = \underbrace{\left(\frac{1}{5} + \frac{1}{5} + \frac{1}{5} + \frac{1}{5} + \frac{1}{5}\right)}_{=1} \times 30 = 30.$$

As it turns out, the unknown x is a number which yields 30, when added to itself 5 times; x satisfies the **unknown-factor equation**,

$$5 \times x = 30,$$

and by the definition of division, x equals $30 \div 5$.

(d) The fact that

$$\frac{1}{5} \times 30 = 6$$

is also consistent with the rules of **multiplicative scaling** (see Chapter 8, pp. 106–108).

Recall: when the multiplier in a product is magnified or reduced, the product is magnified or reduced by the same factor. For example, if the multiplier is reduced by a factor of 5, then the product is reduced by a factor of 5 as well. Start now with the known fact that one is the identity element for multiplication:

$$1 \times 30 = 30.$$

If we reduce the multiplier by a factor of 5 we obtain (by definition!) ONE-FIFTH. By the rules of multiplicative scaling, the product is reduced by a factor of 5 as well, hence,

$$\frac{1}{5} \times 30 = 30 \div 5.$$

31.2.2.2 General fractions multiplied by natural numbers

We are now ready to return to Problem 3:

> There are 30 children in a class. TWO-FIFTHS of them are in the backyard. How many children are in the backyard?

The solution is represented by the product

$$\frac{2}{5} \times 30.$$

We have to determine how many children are TWO-FIFTH of 30 children.

We can solve this problem in two steps:

(a) We already know that ONE-FIFTH of 30 children equals 6 children.
(b) TWO-FIFTHS of a whole is twice ONE-FIFTH of that whole, hence, TWO-FIFTHS of 30 children equal 12 children,

$$\frac{2}{5} \times 30 = 12.$$

Hence, the solution to Problem 3 is:

There are 12 children in the backyard.

What we have done here in disguise, is to use the **associative property** of multiplication; our argument was based on the following steps

$$\frac{2}{5} \times 30 = \left(2 \times \frac{1}{5}\right) \times 30 = 2 \times \underbrace{\left(\frac{1}{5} \times 30\right)}_{=6} = 12.$$

The first equality replaces the fraction TWO-FIFTHS by its definition as 2 times ONE-FIFTH. The second equality applies the associative property.

One may wonder whether the associative property, which was proven correct for natural numbers (Chapter 7, p. 84), holds for products of natural numbers and fractions. The answer is that we have just defined multiplication by a fraction in such a way that ensures that the associative property remains valid. This is a standard practice in mathematics: when a certain system (e.g., a number system) is extended, the operations are extended in a manner that retains their fundamental properties.

Generalizing, let k, m and n be natural numbers. Since a fraction $^m/_n$ of a whole is m times a fraction $^1/_n$ of that whole, letting the whole be k objects, we obtain

$$\frac{m}{n} \times k = m \times \left(\frac{1}{n} \times k\right).$$

We know already that $^1/_n \times k = {}^k/_n$, and we know how to multiply a fraction by a natural number. We obtain the following rule:

For every three natural numbers k, m and n:

$$\frac{m}{n} \times k = m \times \frac{k}{n} = \frac{m \times k}{n}.$$

Please note: since

$$\frac{m}{n} \times k = \frac{m \times k}{n} \qquad \text{and} \qquad k \times \frac{m}{n} = \frac{k \times m}{n},$$

and since the two right-hand sides are equal, once more we have recovered the commutative property, this time, for the product of an arbitrary fraction and a natural number.

31.2.3 Fractions multiplied by fractions

We proceed to examine products in which both operands are fractions. Once again, we progress gradually, starting with the case where both operands are unit fractions.

31.2.3.1 Unit fractions multiplied by unit fractions

Consider a variant of Problem 4, which we call Problem 4a:

> ONE-QUARTER of the campers were assigned to help in decorating the camp. Out of those, ONE-THIRD were asked to hang posters. What fraction of the campers ended up hanging posters?

The solution is represented by the product

$$\frac{1}{3} \times \frac{1}{4}.$$

We have to determine what fraction of a group is ONE-THIRD of ONE-QUARTER of that group.

To illustrate the solution, we model the whole (a group of campers) by a rectangle. Draw a rectangle and color ONE-QUARTER of it orange; to partition the rectangle into quarters, we divide it into 4 equal vertical strips:

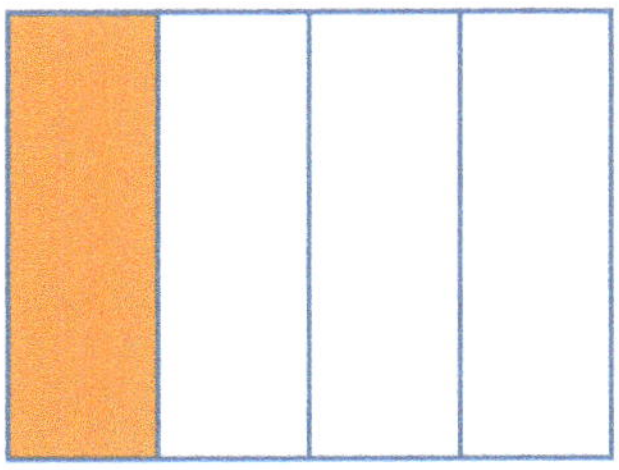

Next, we color green ONE-THIRD of that ONE-QUARTER. We need to take the orange strip, divide it into 3 equal parts, and color one of those parts green. To do so by partitioning the entire rectangle into 3 equal horizontal strips, thus dividing the rectangle into a 3-by-4 rectangular array:

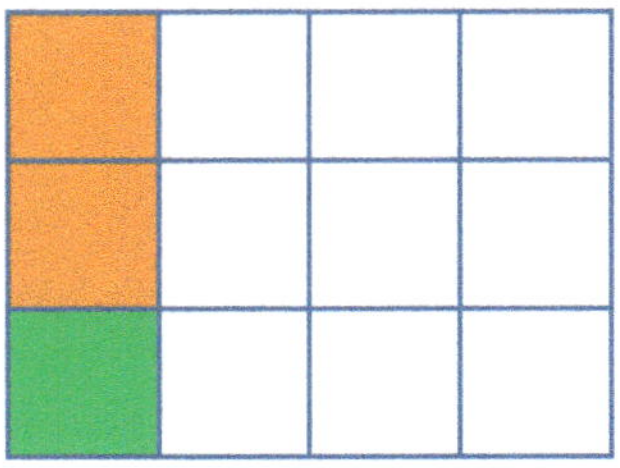

The green region is, by construction, ONE-THIRD of ONE-QUARTER of the whole rectangle. On the other hand, the green region is also a part of the rectangle out of 12 equal parts, i.e., it constitutes ONE-TWELFTH of the rectangle. It follows that

$$\frac{1}{3} \times \frac{1}{4} = \frac{1}{12}.$$

Hence the solution to Problem 4a is:

ONE-TWELFTH of the campers ended up hanging posters.

The merit of the chosen illustration is its generality. Let m and n be natural numbers. Then,

(a) Represent the whole by a rectangle.
(b) Divide the rectangle into n vertical strips of equal width; color one of those strips orange; note that it constitutes a fraction $1/n$ of the rectangle.
(c) Divide the rectangle into m horizontal strips of equal width. This partition divides the orange region into m equal parts. Color one of those parts green.
(d) The green region is, by construction, a fraction $1/m$ of a fraction $1/n$ of the whole rectangle.
(e) At the same time, the green region is a part of the rectangle out of $m \times n$ equal parts, i.e., it constitutes a fraction of $1/(m \times n)$ of the rectangle.

Comparing items (d) and (e), we conclude:

For every two natural numbers m and n:

$$\frac{1}{m} \times \frac{1}{n} = \frac{1}{m \times n}.$$

31.2.3.2 General fractions multiplied by general fractions

We are now in a position to return to Problem 4:

> THREE-QUARTERS of the campers were assigned to help in decorate the camp. Of those, TWO-THIRDS were asked to hang posters. What fraction of the campers ended up hanging posters?

The solution is represented by the product

$$\frac{2}{3} \times \frac{3}{4}.$$

We have to determine what fraction of a group is TWO-THIRDS of THREE-QUARTERS of that group.

We will use the same concrete model as we did for products of unit fractions. We model the whole (a group of campers) by a rectangle. Draw a rectangle and color THREE-QUARTERS of it orange. To partition the rectangle into quarters, we divide it into 4 equal vertical strips:

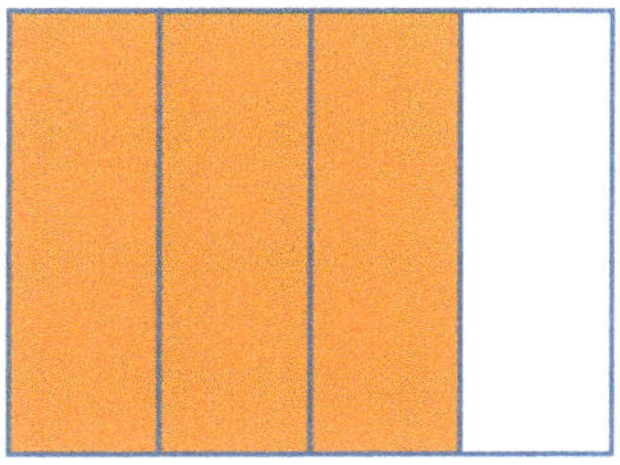

Next, we color TWO-THIRDS of those THREE-QUARTERS green. We need to take the orange region, divide it into 3 equal parts, and color 2 of those parts green. We do so by dividing the entire rectangle into 3 equal horizontal strips, thus partitioning the rectangle into a 3-by-4 rectangular array:

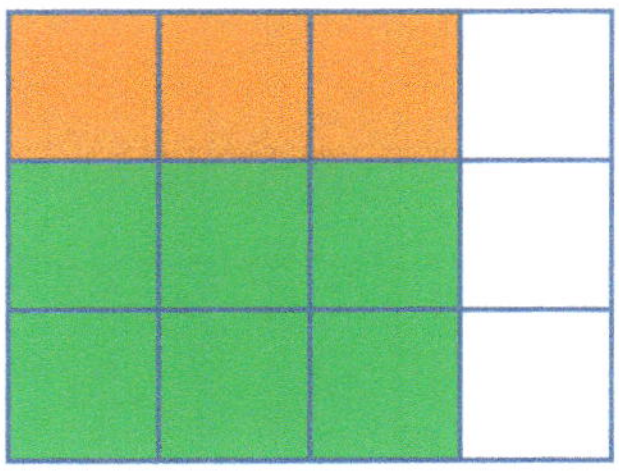

The green region is, by construction, TWO-THIRDS of THREE-QUARTERS of the whole rectangle. On the other hand, the green region comprises 6 out of 12 equal parts, i.e., it constitutes SIX-TWELFTHS of the rectangle. It follows that

$$\frac{2}{3} \times \frac{3}{4} = \frac{6}{12}.$$

Hence the solution to Problem 4 is:

SIX-TWELFTHS of the campers ended up hanging posters.

The resulting fraction SIX-TWELFTHS is reducible; it is equivalent to ONE-HALF. Reducing the final result is good practice because in many cases, fractions are easier to grasp in reduced form (compare your own perception of ONE-HALF versus SIX-TWELFTHS). In some cases, however, 6-out-of-12 may be more meaningful than 1-out-of-2 (e.g., if the class in Problem 4 comprises 12 children). My recommendation is to always seek the representation that is the most meaningful to you.

Our illustration for evaluating products of fractions can be used to derive a general formula. Let k, ℓ, m and n be natural numbers. Then, the product

$$\frac{k}{m} \times \frac{\ell}{n}$$

can be evaluated as follows:

(a) Represent the whole by a rectangle.
(b) Divide the rectangle into n vertical strips of equal width; color ℓ of those strips orange; note that they constitute together a fraction ℓ/n of the rectangle.
(c) Divide the rectangle into m horizontal strips of equal width. This partition divides the orange region into m equal parts. Color k of those parts green.
(d) The green region is, by construction, a fraction k/m of a fraction ℓ/n of the whole rectangle.
(e) At the same time, the green region comprises $k \times \ell$ out of $m \times n$ equal parts, i.e., it constitutes a fraction of $(k \times \ell)/(m \times n)$ of the rectangle.

Comparing Items (d) and (e), we conclude:

For every four natural numbers k, ℓ, m and n:

$$\frac{k}{m} \times \frac{\ell}{n} = \frac{k \times \ell}{m \times n}.$$

As it turns out, to evaluate a product of two fractions we have to multiply numerator by numerator and denominator by denominator. This is a very simple algorithm: unlike addition and subtraction, it does not require finding a common denominator first. Keep in mind, though, the frivolous nature of this simplicity. Procedural proficiency does not guarantee mathematical literacy. To avoid superfluous procedural computations, be sure to provide children with a plethora of word problems challenging their understanding and intuition regarding fraction multiplication.

Other strategies

(a) Hinging on products of unit fractions, we can evaluate products of general fractions by replacing the operands step-by-step. Consider the following sequence of word problem:

(i) ONE-QUARTER of the campers were assigned to help decorate the camp. Of those, ONE-THIRD were asked to hang posters. What fraction of the campers ended up hanging posters?

(ii) THREE-QUARTERS of the campers were assigned to help decorate the camp. Of those, ONE-THIRD were asked to hang posters. What fraction of the campers ended up hanging posters?

(iii) THREE-QUARTERS of the campers were assigned to help decorate the camp. Of those, TWO-THIRDS were asked to hang posters. What fraction of the campers ended up hanging posters?

Problem (i) is the same as Problem 4a. We already know that the solution is ONE-TWELFTH.

In Problem (ii), THREE-QUARTERS of the campers are decorating the camp, compared to ONE-QUARTER in Problem (i). Since there are 3 times as many decorators, and the fraction of poster hangers among decorators hasn't changed, there are also 3 times as many campers hanging posters, namely,

$$3 \times \frac{1}{12} = \frac{3}{12}.$$

Problem (iii) is the same as Problem 4. Now, TWO-THIRDS of the decorators are hanging posters, compared to ONE-THIRD in Problem (ii). Since the total number of decorators hasn't changed, while the fraction of poster hangers among decorators has doubled, then there are twice as many campers hanging posters, namely,

$$2 \times \frac{3}{12} = \frac{6}{12}.$$

(b) Another strategy for obtaining a product of two fractions is based on the rules of **multiplicative scaling** (Chapter 8, pp. 106–108). While somewhat mechanistic, this strategy teaches us to "play with numbers", which has its own merits.

Consider the two factors TWO-THIRDS and THREE-QUARTERS. Since we are more familiar with products of natural numbers, we replace the two fraction-valued operands by natural numbers by magnifying them appropriately:

$$3 \times \frac{2}{3} = 2 \qquad \text{and} \qquad 4 \times \frac{3}{4} = 3.$$

Of course, we have to remember that we have magnified one factor by 3 and the other by 4. By the rules of multiplicative scaling, the product is also magnified by these two factors, i.e., the product is magnified by a factor of 12.

Next, evaluate the product of the two natural number thus obtained:

$$2 \times 3 = 6.$$

Finally, to obtain the product of the two original operands, we have to reduce this product by a factor of 12, obtaining

$$6 \div 12 = \frac{6}{12}.$$

This strategy can be illustrated by the following scheme:

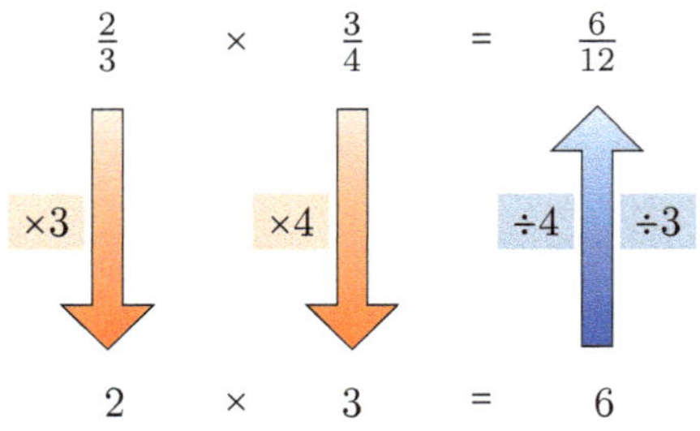

31.3 Multiplicative inverses

Consider the following products:

$$2 \times \frac{1}{2}, \qquad \frac{2}{3} \times \frac{3}{2} \qquad \text{and} \qquad \frac{5}{7} \times \frac{7}{5}.$$

Now that we can multiply any two rational numbers, we evaluate all three products:

$$2 \times \frac{1}{2} = \frac{2}{2} = 1$$
$$\frac{2}{3} \times \frac{3}{2} = \frac{2 \times 3}{3 \times 2} = \frac{6}{6} = 1$$
$$\frac{5}{7} \times \frac{7}{5} = \frac{5 \times 7}{7 \times 5} = \frac{35}{35} = 1.$$

That is, all three products equal 1.

This observation leads to the following definition:

Definition 31.1. A pair of rational number a and b satisfying

$$a \times b = 1$$

are called **multiplicative inverses**, or **reciprocal**. Given a rational number a, its multiplicative inverse is the rational number solving the unknown-factor equation

$$\boxed{?} \times a = 1.$$

Thus, 2 and ONE-HALF are multiplicative inverses, and so are the pairs of numbers TWO-THIRDS and THREE-HALVES, and FIVE-SEVENTHS and SEVEN-FIFTHS. The only number having no multiplicative inverse is zero, because there exists no number a satisfying the unknown-factor equation

$$a \times 0 = 1$$

(see Chapter 7, pp. 110–112).

Given a (non-zero) rational number, finding its multiplicative inverse is quite easy. If this number is a fraction, you only need to interchange numerator and denominator: for every natural numbers m and n,

$$\frac{n}{m} \quad \text{is the multiplicative inverse of} \quad \frac{m}{n}.$$

In the particular case where $m = 1$, recalling that $^n/_1 = n$, we have that

$$n \quad \text{is the multiplicative inverse of} \quad \frac{1}{n}.$$

At this stage of the book, you have surely guessed that defining a new concept—a multiplicative inverse—and learning how to obtain it using a procedural algorithm is not what we are aiming at. It is more important to understand the meaning of multiplicative inverses, and how to use them.

Uses of multiplicative inverses will be described in the next chapter, in the context of fraction division. As for the meaning of a multiplicative inverse, consider an example:

Which number of the multiplicative inverse of THREE-QUARTERS?

By the definition of multiplicative inverses, we are seeking a rational number satisfying the unknown-factor equation

$$\boxed{?} \times \frac{3}{4} = 1.$$

In other words, *the multiplicative inverse of a fraction is the amount of that fraction needed to form a whole.*

Let's illustrate the task of finding a multiplicative inverse modeling the whole by a disc. THREE-QUARTERS of a disc are displayed below:

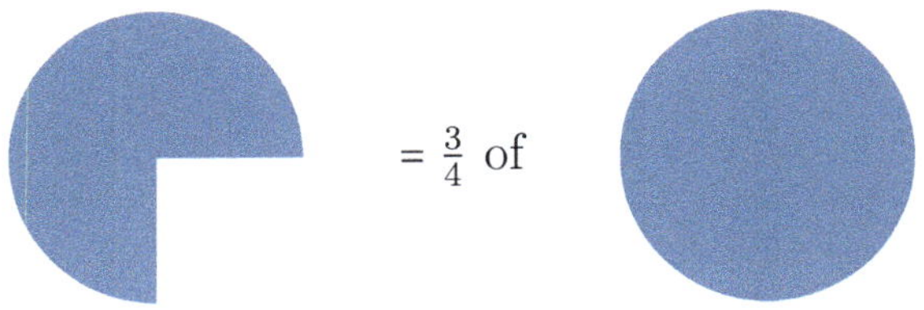

To find what fraction of THREE-QUARTERS of a whole equals a whole, we rewind the process generating THREE-QUARTERS of a whole. We start by separating the THREE-QUARTERS into 3 distinct quarters:

Each quarter of the whole is ONE-THIRD of THREE-QUARTERS of that whole, namely

 $= \frac{1}{3}$ of 

The whole equals 4 times its quarter, hence

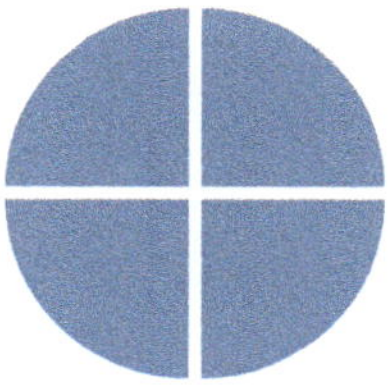 $= \frac{4}{3}$ of 

That is, FOUR-THIRDS is the fraction inverse to THREE-QUARTERS.

A similar perspective is obtained by stating that since THREE-QUARTERS are obtained from the whole by partitioning it into 4 equal parts and taking 3 of the parts, then the whole is obtained from its THREE-QUARTERS by first partitioning them into 3 equal parts and then taking 4 such parts.

31.4 Is multiplication a magnifying operation?

Children (and adults) often perceive multiplication as a magnifying operation. In this section, we will elucidate what is meant when saying that multiplication magnifies, and we will see in which instances this statement is correct.

Let's start with an example. Consider the product:

$$7 \times 8.$$

In what sense does this multiplication magnify? What is being magnified?

One might argue that the product, 56, is larger that the two factors, 7 and 8, and this is what we mean when we say that multiplication magnifies. This interpretation is not satisfactory, because it is not clear in which sense a product magnifies its factors.

A better interpretation views multiplication as a **scaling** operation. The product 7×8 can be viewed as the outcome of scaling, or magnifying the multiplicand, 8, by a factor of 7; in fact, this is precisely the interpretation of multiplication as repeated addition. If other words, we identify the product 7×8 as the action of the numerical operator SEVEN OF on the input 8. Note the asymmetric role of the two factors: one is interpreted as a number and the other is interpreted as a numerical operator.

Viewed from this angle, we can clarify the concept of a magnifying operation. The action of the numerical operator SEVEN OF returns seven copies of its input, hence it is a magnifying operation. Rather than saying that multiplication is a magnifying operation, we should say that the scaling operation modeled by multiplication is a magnifying operation.

But is this alway so? Is any multiplicative scaling a magnifying operation? No. If the scaling factor is greater than one, then the output of the numerical operator is greater than its input. If, however, the scaling factor is less than one, then the output of the numerical operator is less than its input. The action of numerical operators such as ONE-HALF OF or TWO-THIRDS OF is actually reducing.

The fact that not every scaling is magnifying is counter-intuitive to many children. Children are often confused when evaluating products such that

$$\frac{3}{4} \times 12 = 9,$$

discovering that the product, 9, is less than the multiplicand, 12. As always, the remedy to confusion is emphasizing meaning, and decoding number sentences using concrete means.

To conclude, the scaling properties of multiplication are:

(a) Multiplication by a number greater than 1 is magnifying.
(b) Multiplication by 1 does not change the input.
(c) Multiplication by a number less than 1 is reducing.

Mathematical problems

Problem 31.1. Consider the following word problems:

(i) Eran organizes a party. He plans to bake one cake whose recipe requires 4 pounds of flour. When he learns that there are going to be 3 times as many guests as planned, he decides to bake 3 such cakes. How many pounds of flour will he need?

(ii) Eran organizes a party. He plans to bake one cake whose recipe requires FOUR-FIFTHS of a pound of flour. When he learns that there are going to be 3 times as many guests as planned, he decides to bake 3 such cakes. How many pounds of flour will he need?

(iii) Eran organizes a party. He plans to bake one cake whose recipe requires 4 pounds of flour. When he learns that the number of guests is going to be ONE-THIRD the number he was expecting, he decides to bake a cake whose size is ONE-THIRD the size he was planning to bake. How many pounds of flour will he need?

(iv) Eran organizes a party. He plans to bake one cake whose recipe requires FOUR-FIFTHS of a pound of flour. When he learns that the number of guests is going to be ONE-THIRD the number he was expecting, he decides to bake a cake whose size is ONE-THIRD the size he was planning to bake. How many pounds of flour will he need?

(a) Write an arithmetic expression representing each solution and evaluate it.
(b) What is common to all four questions? Are they good illustrations of the meaning of rational number multiplication?

Problem 31.2. A birthday cake was divided among 6 children. First, Roee ate ONE-SIXTH of the cake. Then, Iddo ate ONE-FIFTH of the portion left by Roee. Yuval ate ONE-QUARTER of the portion left by Iddo and Mai ate ONE-THIRD of the portion left by Yuval. Finally, Matan and Sivan equally divided the part Mai left.

(a) Who got the largest portion?
(b) Write an arithmetic expression for the fraction of cake eaten by Mai.

Problem 31.3. Consider the following sequence:

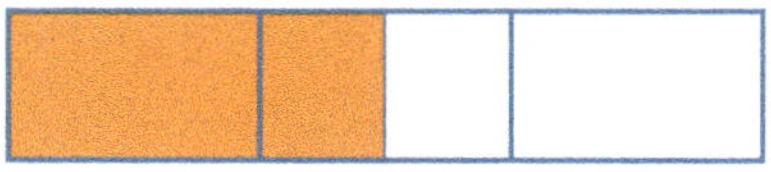

$\frac{1}{3} + \frac{1}{2} \times \frac{1}{3} = \frac{1}{2}$

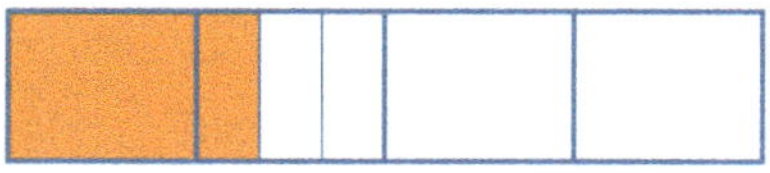

$\frac{1}{4} + \frac{1}{3} \times \frac{1}{4} = \frac{1}{3}$

$\frac{1}{5} + \frac{1}{4} \times \frac{1}{5} = \frac{1}{4}$

(a) How does each number sentence relate to the figure to its left?
(b) Draw the next figure in the sequence along with the appropriate arithmetic expression.

Problem 31.4. A chicken and a half lays an egg and a half every day and a half. How many eggs are laid by 3 chickens in 3 days? (This riddle is famous; relate its solution to the content of this chapter.)

Problem 31.5. Evaluate the products

$$2\frac{3}{4} \times 5$$
$$6\frac{1}{3} \times \frac{5}{8}$$
$$2\frac{1}{2} \times 3\frac{1}{2}.$$

(a) By first transforming the mixed numbers into improper fractions.
(b) By using the distributive property.

Which method is more convenient?

Problem 31.6. Place the numbers 3, 4, 5 and 6 in the following template

$$\frac{\square}{\square} \times \frac{\square}{\square}$$

to obtain (a) a maximal product (b) a minimal product.

Pedagogical problems

Problem 31.7. Use an area model to illustrate the product of the following improper fractions:

$$\frac{3}{2} \times \frac{5}{4}.$$

Problem 31.8. Ms Merz asked her class to find a word problem whose solution is given by the arithmetic expression

$$\frac{1}{5} \times \frac{1}{3}.$$

Nadav's suggestion was: ONE-FIFTH of the children brought ONE-THIRD of a gallon of water each for the field trip. What was the total number of gallons of water they brought?

Is Nadav's answer correct? Justify your answer.

Chapter 32

Fraction Division

Division is considered at large the most difficult of all arithmetic operations. Fraction division combines the difficulties accompanying division with the difficulties accompanying the internalization of the concept of fractions. From my own experience, most adults don't even understand the *meaning* of dividing fractions. Many would not be able to find a concrete problem justifying the evaluation of the quotient

$$1\frac{1}{2} \div \frac{2}{3}.$$

The absence of meaning results in a procedural *know how* approach. Regrettably, most of class practice focuses on the evaluation of quotients of fractions, rather than providing contexts in which fraction division occurs. Every child in the United States knows the slogan:

> Ours is not to reason why
> Just invert and multiply!

This slogan encourages children to focus on procedure rather than on meaning. The main goal of this chapter is to counteract the damages that this slogan has inflicted on generations of learners. Hopefully, by the end of this chapter you will have a wealth of settings calling for fraction division. Moreover, you will understand *why* a quotient of fractions is evaluated by *inverting and multiplying.*

Hurdles:

(a) Children don't understand the meaning of fraction division and are content with knowing how to evaluate products.
(b) Children do not regard whole number division and fraction division as the same operation.
(c) Children have a misconception whereby division is always a reducing operation.

Remedies:

(a) Practice fraction division in the context of concrete word problems; use both partitive division and quotative division settings.
(b) Study division problems in which the operands can be interchangeably whole numbers or fractions; this will create a connection between the division operation in both contexts.
(c) Present division as an unknown-factor problem; this interpretation is sensible for both whole numbers and fractions.
(d) Use visual aids to demonstrate fraction division.
(e) Use concrete models to demonstrate why division by a fraction is not reducing, but rather magnifying.

32.1 Division revisited

Th division operation has two principal interpretations, both ensuing from being an operation inverse to multiplication. For example, the quotient

$$12 \div 3$$

may represent a solution to two types of problems (Chapter 8, pp. 97–100):

(a) **Partitive division**: 12 apples were divided evenly among 3 children. How many apples did each child get?

(b) **Quotative division**: 12 apples were bagged up, 3 apples in each bag. How many bags were required?

In partitive division, there is a given number of objects (12 apples), shared equally into a given number of sets (3, as the number of children). We look for the number of objects in each set, or the size of each **part**. This problem can be viewed as an **unknown-multiplicand** equation:

$$3 \times \boxed{?} = 12.$$

In quotative division, there is a given number of objects (12 apples), partitioned into equal sets. We know the number of objects in each set, or the size of each **quota** (3 apples). This time we look for the number of sets. This problem can be viewed as an **unknown-multiplier** equation:

$$\boxed{?} \times 3 = 12.$$

Take now the arithmetic expression

$$1\frac{1}{2} \div \frac{2}{3}.$$

In an attempt to make sense of it, let's take the same two words problems, replacing the integer operands, 12 and 3 by the fractional operands, ONE-AND-A-HALF and TWO-THIRDS:

(a) **Partitive division**: ONE-AND-A-HALF apples were divided evenly among TWO-THIRDS of a child. How many apples did each child get?

(b) **Quotative division**: ONE-AND-A-HALF apples were bagged up, TWO-THIRDS of an apple in each bag. How many bags were required?

The first question is bizarre; how can we interpret TWO-THIRDS of a child, and what does *each child* mean in this context? The second question is more sensible; one could cut up the apples and place TWO-THIRDS of an apple in each bag. Yet, there remains a nagging issue: how should be deal with the remainder, if we are left with less than TWO-THIRDS of an apple? It is precisely this sort of uncertainty that lead many learners to the sad conclusion that fraction division is meaningless. From here, there is a short path to *just invert and multiply.*

This confusion is the result of a lack of concrete contexts for division. Understanding fraction division hinges above all on understanding whole number division, where by *understanding*, I mean being able to devise an abundance of disparate situations modeled by division. In the next section, we will present a selection of concrete settings modeled by division. Each of those settings will be shown, further below, to constitute a natural setting for fraction division.

32.1.1 Quotients of natural numbers

Consider the arithmetic expression:

$$12 \div 3.$$

We make this expression concrete by formulating six word problems, whose solutions are modeled by this quotient.

Problem 1

> 12 quarts of water are required for watering 3 rows of flowers. How many quarts of water are required for watering one row?

This problem is an instance of **partitive division**. The following rephrasing casts the problem into the prevalent pattern of partitive division:

> 12 quarts of water are divided equally among 3 rows. How many quarts of water does each row get?

Solution: 4 quarts of water are required for each row.

Problem 2

> Suitcase A weighs 3 times as much as Suitcase B. How much does Suitcase B weigh if Suitcase A weighs 12 pounds?

This problem is also an instance of **partitive division**, even though this is not as apparent as in the previous problem. The following rephrasing emphasizes the partitive nature of this problem:

> If we divide a 12 pound weight into 3 equal parts, each part weighs as much as Suitcase B. How much does Suitcase B weigh?

Solution: Suitcase B weighs 4 pounds.

Problem 3

> A farmer plows 3 fields in 12 days. How many days does it take him to plow one field at that pace?

This is also an instance of **partitive division**, as apparent from the following rephrasing:

> A farmer spends 12 days plowing 3 fields. Each field takes the same amount of time to plow. How many days does it take him to plow each field?

Solution: It takes him 4 days to plow one field.

Problem 4

> We need 3 pounds of flour to bake one loaf of bread. How many loaves can we bake with 12 pounds of flour?

This problem is an instance of **quotative division**, as apparent from the following rephrasing:

> If 12 pounds of flour are distributed among loaves of bread, 3 pounds to each loaf, how many loaves of bread are there?

Solution: 4 loaves of bread can be baked.

Problem 5

> The area of Room A is 12 sq-meters. The area of Room B is 3 sq-meters. How many times larger is the area of Room A than the area of Room B?

This problem is also an instance of **quotative division**, as apparent from the following rephrasing:

> A 12 sq-meter room is partitioned into smaller rooms, 3 sq-meters each. How many smaller rooms will we have?

Solution: Room A is 4 times as large as Room B.

Problem 6

> The area of a rectangle is 12 sq-feet. The length of one of its sides is 3 feet. What is the length of the adjacent side?

This problem neither matches partitive nor quotative division. One would recognize an **unknown-factor problem**, but the symmetric roles of two adjacent sides of a rectangle do not distinguish the multiplier from the multiplicand. One could match to this problem either unknown-factor equation

$$3 \times \boxed{?} = 12 \qquad \text{or} \qquad \boxed{?} \times 3 = 12.$$

Solution: The length of the adjacent side is 4 feet.

32.1.2 Quotients of fractions

Now that we have seen examples of six word problems modeling division, let's get back to the quotient

$$1\frac{1}{2} \div \frac{2}{3}$$

and substitute the operands ONE-AND-A-HALF and TWO-THIRDS in place of the operands 12 and 3 in each of the six word problems:

	$12 \div 3$	$1\frac{1}{2} \div \frac{2}{3}$
Partitive	12 quarts of water are required for watering 3 rows of flowers. How many quarts of water are required for watering one row?	ONE-AND-A-HALF quarts of water are required for watering TWO-THIRDS of a row of flowers. How many quarts of water are required for watering one row?
Partitive	Suitcase A weighs 3 times as much as Suitcase B. How much does Suitcase B weigh if Suitcase A weighs 12 pounds?	Suitcase A weighs TWO-THIRDS of Suitcase B. How much does Suitcase B weigh if Suitcase A weighs ONE-AND-A-HALF pounds?
Partitive	A farmer plows 3 fields in 12 days. How many days does it take him to plow one field at that pace?	A farmer plows TWO-THIRDS of a field in ONE-AND-A-HALF days. How many days does it take him to plow one field at that pace?
Quotative	We need 3 pounds of flour to bake one loaf of bread. How many loaves can we bake with 12 pounds of flour?	We need TWO-THIRDS of a pound of flour to bake one loaf of bread. How many loaves can we bake with ONE-AND-A-HALF pounds of flour?

Quotative	The area of Room A is 12 sq-meters. The area of Room B is 3 sq-meters. How many times larger is the area of Room A than the area of Room B?	The area of Room A is ONE-AND-A-HALF sq-meters. The area of Room B is TWO-THIRDS sq-meters. How many times larger is the area of Room A than the area of Room B?
N/A	The area of a rectangle is 12 sq-feet. The length of one of its sides is 3 feet. What is the length of the adjacent side?	The area of a rectangle is ONE-AND-A-HALF sq-feet. The length of one of its sides is TWO-THIRDS of a foot. What is the length of the adjacent side?

Carefully read each of the six word problems on the left column along with its counterpart on the right column. Unlike the failed attempt to divide ONE-AND-A-HALF apples to TWO-THIRDS of a child, each of those problems is sensible even when the operands are fractions. We have thus obtained six disparate settings calling for fraction division. I strongly encourage you to try solving these word problems with your own methods prior to learning systematic methods for evaluating quotients of fractions.

32.1.3 Division is inverse to multiplication

In Chapter 8, we identified division as an operation inverse to multiplication—the value of a quotient is the solution of an unknown-factor equation. In Chapter 31, p. 234, we interpreted products (of natural numbers and fractions alike) as a serial linking of two **numerical operators**. Combining the two, we will now interpret division in terms of an *unknown numerical operator*.

This idea is best demonstrated with a specific example in mind. Consider the word problem:

> A group comprises 40 children. TWO-FIFTHS of them play chess. How many children in the group play chess?

Using numerical operators, the chess-playing children are the output of the following serial network:

The group of chess-playing children is the output of the TWO-FIFTHS OF operator. The input of the latter is the output of the FORTY OF operator, which in turn, takes one child for input. The solution to the problem entails finding a single numerical operator whose action is the same as the serial action of FORTY OF followed by TWO-FIFTHS OF. When we say that

$$\frac{2}{5} \times 40 = 16,$$

we mean that

Division arises when one of the factors in a product is unknown. Since the multiplicand and the multiplier play different roles, we can match the word problem to two inverse problems, each exhibiting a different unknown:

(a) There are 40 children in a group. 16 of them play chess. Which fraction of the children in the group plays chess?
(b) 16 children, who constitute TWO-FIFTHS of a group, play chess. How many children are in the group?

The solution of the first question can be modeled by an equation of networks of numerical operators:

We seek what fraction of 40 equals 16. The solution is the value of the arithmetic expression

$$16 \div 40.$$

The solution of the second question can be modeled by the following equation:

We seek a number that TWO-FIFTHS of it equal 16. The solution is the value of the arithmetic expression

$$16 \div \frac{2}{5}.$$

32.2 Evaluating quotients

32.2.1 Fractions divided by whole numbers

Consider the word problem:

> It takes TWO-THIRDS of a quart of water to water 5 flowerpots. How many quarts of water are required to water 1 flowerpot?

This is a partitive division problem. TWO-THIRDS of a quart of water are allocated evenly among 5 flowerpots. We seek how many quarts of water are allocated to each flowerpot. The corresponding unknown-multiplicand equation is

$$5 \times \boxed{?} = \frac{2}{3}.$$

The value of the unknown is represented by the quotient

$$\frac{2}{3} \div 5.$$

We evaluate this quotient using five different strategies:

(a) **A concrete model**: Model the whole by a rectangle, as in Chapter 31. First, color TWO-THIRDS of that rectangle orange:

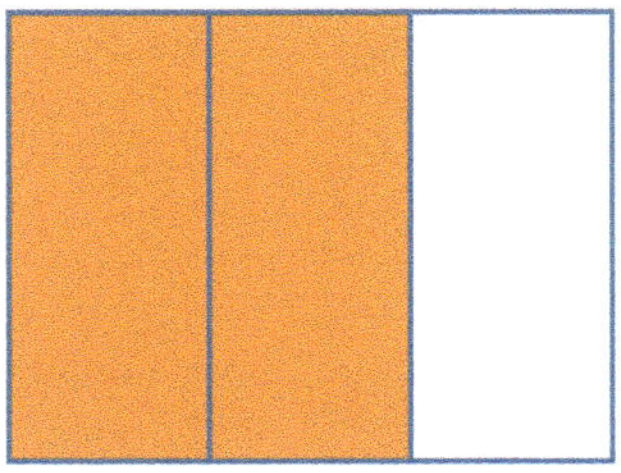

We need to divide those TWO-THIRDS of a rectangle by 5. To accomplish this, we partition the *entire* rectangle into 5 horizontal strips, and color the orange region in one of those strips green.

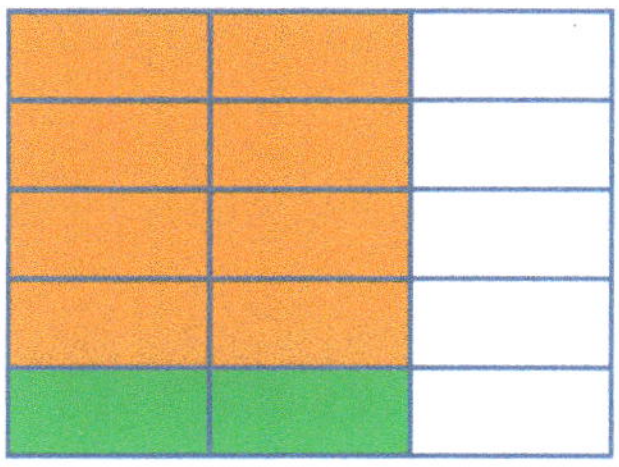

The green region, which models TWO-FIFTHS of a whole divided by 5, comprises 2 rectangles out of 15, hence, it constitutes TWO-FIFTEENTHS of the whole:

$$\frac{2}{3} \div 5 = \frac{2}{15}.$$

Solution: Each flowerpot requires TWO-FIFTEENTHS of a quart of water.

(b) **Gradual replacement of the operands**: Suppose momentarily that we only need ONE-THIRD of a quart of water to water 5 flowerpots. ONE-THIRD of a quart is a portion obtained by dividing 1 quart into 3 equal parts. If we divide ONE-THIRD of a quart into 5 parts of equal size, then each portion is of the same size as when a whole quart is divided into 15 equal parts. That is:

$$\frac{1}{3} \div 5 = \frac{1}{15}.$$

However, the amount of water needed to water 5 flowerpots is TWO-THIRDS of a quart, which is twice as much as ONE-THIRD of a quart. Therefore, the amount of water needed to water 1 flowerpots is twice as much as it would have been if 5 flowerpots required only ONE-THIRD of a quart. It follows that,

$$\frac{2}{3} \div 5 = 2 \times \frac{1}{15} = \frac{2}{15}.$$

(c) **Multiplication by a unit fraction**: In Chapter 31 we discovered that multiplication by $^1/_n$ amounts to division by n. Using this equivalence in the reverse direction and applying the algorithm for products of fractions,

$$\frac{2}{3} \div 5 = \frac{1}{5} \times \frac{2}{3} = \frac{1 \times 2}{5 \times 3} = \frac{2}{15}.$$

This method can be generalized:

> For every three natural numbers k, m and n
>
> $$\frac{m}{n} \div k = \frac{1}{k} \times \frac{m}{n} = \frac{m}{k \times n}.$$

(d) **Using multiplicative scaling**: As learned on p. 184, the fraction TWO-THIRDS is the value of the quotient $2 \div 3$. Thus,

$$\frac{2}{3} \div 5 = (2 \div 3) \div 5.$$

The rules of **multiplicative scaling** (Chapter 8, p. 107) imply that for every three numbers a, b and c,

$$(a \div b) \div c = a \div (b \times c).$$

Thus,

$$(2 \div 3) \div 5 = 2 \div (3 \times 5) = 2 \div 15.$$

Since $2 \div 15$ equals TWO-FIFTEENTHS, we obtain once again

$$\frac{2}{3} \div 5 = \frac{2}{15}.$$

(e) **Multiplying by the inverse and using the associative property**: We are seeking to solve the unknown-multiplicand equation

$$5 \times \boxed{?} = \frac{2}{3}.$$

Start with the identity connecting the divisor 5 to its multiplicative inverse,

$$5 \times \frac{1}{5} = 1.$$

Next, multiply both sides by TWO-THIRDS,

$$\left(5 \times \frac{1}{5}\right) \times \frac{2}{3} = 1 \times \frac{2}{3} = \frac{2}{3}.$$

Apply now the associative property on the left-hand side,

$$5 \times \left(\frac{1}{5} \times \frac{2}{3}\right) = \frac{2}{3}.$$

Evaluating the product in parentheses,

$$5 \times \frac{2}{15} = \frac{2}{3}.$$

The fraction TWO-FIFTEENTHS solves the unknown-multiplicand equation.

Methods (c) through (e) are based on what I call "playing with numbers". I do not recommend them at initial learning stages, where more concrete, hands-on approaches are needed for acquiring conceptual understanding. Yet, "playing with numbers" is important, as it is very often a good problem-solving strategy. Moreover, the manipulation of numbers is an important prelude to the manipulation of algebraic expressions in higher grades.

32.2.2 Whole numbers divided by fractions

Consider the word problem:

> To bake a loaf of bread TWO-THIRDS of a pound of flour are required. How many loaves can be baked with 5 pounds of flour?

This is a quotative division problem. We seek how many quotas of TWO-THIRDS of a pound are contained in 5 pounds. The corresponding unknown-multiplier equation is

$$\boxed{?} \times \frac{2}{3} = 5.$$

The value of the unknown is represented by the quotient

$$5 \div \frac{2}{3}.$$

We evaluate this quotient using four different strategies:

(a) **A concrete model**: Model the whole by a disc. We seek how many TWO-THIRDS of a disc are contained in 5 discs. To this end, take 5 discs, partition each into 3 equals sectors, and using various colors, color non-overlapping regions whose sizes are TWO-THIRDS of a disc:

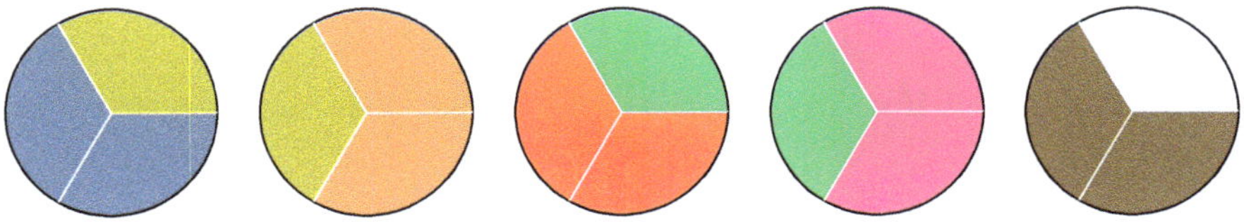

We find that the divisor—TWO-THIRDS of a disc—is contained 7 times in the dividend—5 discs. But there is a **remainder** of ONE-THIRD of a disc that has not yet been colored.

To complete the division process, we have to find out how many times are TWO-THIRDS of a disc contained in ONE-THIRD of a disc. ONE-THIRD of a disc is ONE-HALF of TWO-THIRDS of a disc, hence the number of times that the TWO-THIRDS of a disc are contained in the 5 discs is SEVEN-AND-A-HALF, namely,

$$5 \div \frac{2}{3} = 7\frac{1}{2}.$$

Solution: One can make SEVEN-AND-A-HALF loaves of bread with 5 pounds of flour.

(b) **Gradual replacement of the operands**: Suppose momentarily that ONE-THIRD of a pound of flour was required to bake a loaf of bread. It follows that one pound of flour makes 3 loaves, and 5 pounds make 5 times as many—15 loaves. Hence,

$$5 \div \frac{1}{3} = 15.$$

However, the amount of flour needed to make a loaf of bread is TWO-THIRDS of a pound, which is twice as much as ONE-THIRD of a pound. Therefore, the number of loaves one can make with 5 pounds equals the number of loaves one could have made if only ONE-THIRD of a pound was needed per loaf, divided by 2:

$$5 \div \frac{2}{3} = 15 \div 2 = \frac{15}{2}.$$

Turning this improper fraction into a mixed number we obtain a quotient of SEVEN-AND-A-HALF.

(c) **Using multiplicative scaling**: Start with the trivial identity,

$$5 \div 1 = 5.$$

By the rules of multiplicative scaling (Chapter 8, p. 107), if we reduce the divisor 3-fold, then the quotient is magnified 3-fold,

$$5 \div (1 \div 3) = 3 \times 5 = 15,$$

which amount to

$$5 \div \frac{1}{3} = 15.$$

Conversely, if we magnify the divisor 2-fold, then the quotient is reduced 2-fold,

$$5 \div \left(2 \times \frac{1}{3}\right) = 15 \div 2.$$

Since 2 times ONE-THIRD equal TWO-THIRDS and since $15 \div 2$ equals FIFTEEN-HALVES, we obtain once again

$$5 \div \frac{2}{3} = \frac{15}{2}.$$

(d) **Multiplying by the inverse and using the associative property**: We are seeking to solve the unknown-multiplier equation

$$\boxed{?} \times \frac{2}{3} = 5.$$

The multiplicative inverse of TWO-THIRDS is THREE-HALVES,

$$\frac{3}{2} \times \frac{2}{3} = 1.$$

Therefore,

$$5 \times \underbrace{\left(\frac{3}{2} \times \frac{2}{3}\right)}_{=1} = 5.$$

Next, apply the associative property of multiplication on the left-half side,

$$\left(5 \times \frac{3}{2}\right) \times \frac{2}{3} = 5.$$

Evaluating the product in parentheses,

$$\frac{15}{2} \times \frac{2}{3} = 5.$$

The fraction FIFTEENTH-HALVES solves the unknown-multiplier equation.

This last method, based on the multiplicative inverse, can be generalized: Let k, m and n be natural numbers, and consider the quotient

$$k \div \frac{m}{n},$$

whose value solves the unknown-multiplier equation,

$$\boxed{?} \times \frac{m}{n} = k.$$

Since

$$\frac{n}{m} \times \frac{m}{n} = 1,$$

it follows that

$$k \times \underbrace{\left(\frac{n}{m} \times \frac{m}{n}\right)}_{=1} = k.$$

Applying the associative property of multiplication on the left-half side,

$$\left(k \times \frac{n}{m}\right) \times \frac{m}{n} = k.$$

Evaluating the product in parentheses,

$$\frac{k \times n}{m} \times \frac{m}{n} = k.$$

Comparing this equation to the unknown-multiplier equation, we deduce that

> For every three natural numbers k, m and n
>
> $$k \div \frac{m}{n} = \frac{k \times n}{m}.$$

32.2.3 Fractions divided by fractions

Consider the word problem:

> To bake a loaf of bread TWO-THIRDS of a pound of flour are required. How many loaves can be baked with FOUR-FIFTHS of a pound of flour?

This problem is identical to the previous one, with the exception of one numerical change. This time, we seek how many quotas of TWO-THIRDS of a pound are contained in FOUR-FIFTHS of a pound. The corresponding unknown-multiplier equation is

$$\boxed{?} \times \frac{2}{3} = \frac{4}{5}.$$

The value of the unknown is represented by a quotient of two fractions,

$$\frac{4}{5} \div \frac{2}{3}.$$

We evaluate this quotient using the four strategies presented in the previous section. There will be some repetition. However, note the subtle differences, and remember our goal of keeping away from the slippery slope of *just invert and multiply.*

(a) **A concrete model**: We seek how many TWO-THIRDS of a disc are contained in FOUR-FIFTHS of a disc.

Draw FOUR-FIFTHS of a disc, and next to it TWO-THIRDS of a disc

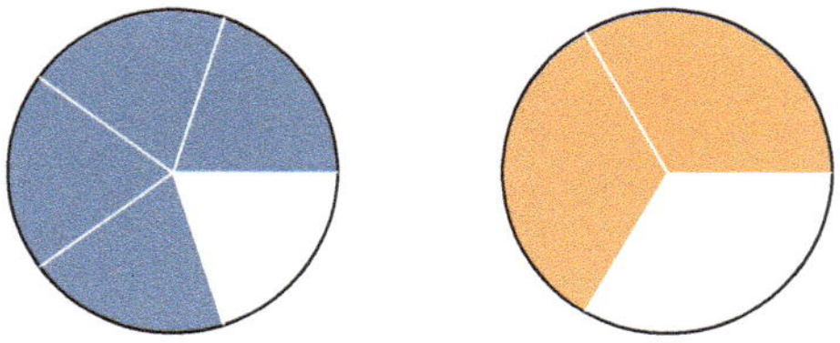

It takes more than TWO-THIRDS of a disc to make FOUR-FIFTHS of a disc, hence the quotient is expected to be larger than one. However, it is not clear how to quantify that quotient.

The solution hinges on using a common denominator. Since 20 is a common multiple of 5 and 4, we expand the dividend FOUR-FIFTHS by 3 to get the equivalent fraction TWELVE-FIFTEENTHS. In our concrete model, it entails the partition of each FIFTH of the disc into 3 equal sectors. Similarly, we expand the divisor TWO-THIRDS by 5 to get the equivalent fraction TEN-FIFTEENTHS. In our concrete model, it entails the partition of each THIRD of the disc into 5 equal sectors:

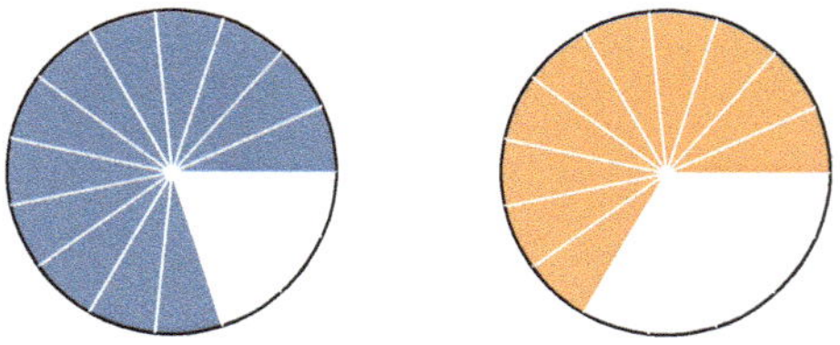

Next, we decompose the dividend in a way reminiscent of the **chunking** method:

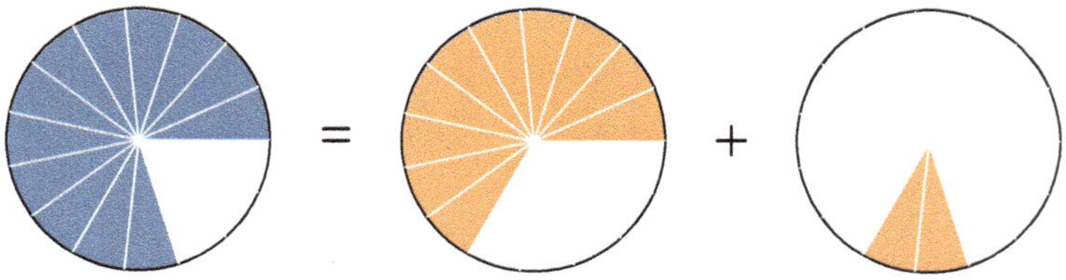

Consider the sum on the right-hand side: the first addend equals the divisor. The second addend can be viewed as a part of the divisor. Since the divisor comprises 10 colored sectors of equal size, and the second addend comprises 2 sectors of identical size, the second addend equals TWO-TENTHS of the divisor.

Hence, the divisor is contained in the dividend ONE-AND-TWO-TENTHS times, namely

$$\frac{4}{5} \div \frac{2}{3} = 1\frac{2}{10}.$$

The fraction TWO-TENTHS can be further reduced into ONE-FIFTH.

Solution: One can make ONE-AND-ONE-FIFTH loaves of bread with FOUR-FIFTHS of a pound of flour.

(b) **Gradual replacement of the operands**: Suppose momentarily that ONE-THIRD of a pound of flour was required to bake a loaf of bread. It would follow that one pound of flour makes 3 loaves. Then, FOUR-FIFTHS of a pound would only make FOUR-FIFTHS of 3 loaves, namely TWELVE-FIFTHS of a loaf. Hence,

$$\frac{4}{5} \div \frac{1}{3} = \frac{12}{5}.$$

The amount of flour needed to make a loaf of bread is however TWO-THIRDS of a pound, which is twice as much as ONE-THIRD of a pound. Therefore, the number of loaves one can make with FOUR-FIFTHS of a pound equals the number of loaves one could make if only ONE-THIRD of a pound was needed per loaf, divided by 2:

$$\frac{4}{5} \div \frac{2}{3} = \frac{12}{5} \div 2 = \frac{12}{10}.$$

Turning this improper fraction into a mixed number we obtain a quotient of ONE-AND-TWO-TENTHS, and after further simplification, ONE-AND-ONE-FIFTH.

(c) **Using multiplicative scaling**: Start with the trivial identity,

$$\frac{4}{5} \div 1 = \frac{4}{5}.$$

By the rules of multiplicative scaling (Chapter 8, p. 107), if we reduce the divisor 3-fold, the quotient is magnified 3-fold,

$$\frac{4}{5} \div (1 \div 3) = 3 \times \frac{4}{5} = \frac{12}{5}.$$

Since $1 \div 3$ equals ONE-THIRD,

$$\frac{4}{5} \div \frac{1}{3} = \frac{12}{5}.$$

Conversely, if we magnify the divisor 2-fold, the quotient is reduced 2-fold,

$$\frac{4}{5} \div \left(2 \times \frac{1}{3}\right) = \frac{12}{5} \div 2 = \frac{12}{10}.$$

Since 2 times ONE-THIRD equals TWO-THIRDS, we obtain once again

$$\frac{4}{5} \div \frac{2}{3} = \frac{12}{10}.$$

(d) **Multiplying by the inverse and using the associative property**: We are seeking to solve the unknown-multiplier equation

$$\boxed{?} \times \frac{2}{3} = \frac{4}{5}.$$

The number inverse to TWO-THIRDS is THREE-HALVES,

$$\frac{3}{2} \times \frac{2}{3} = 1.$$

Therefore,

$$\frac{4}{5} \times \underbrace{\left(\frac{3}{2} \times \frac{2}{3}\right)}_{=1} = \frac{4}{5}.$$

Next, apply the associative property of multiplication on the left-half side,

$$\left(\frac{4}{5} \times \frac{3}{2}\right) \times \frac{2}{3} = \frac{4}{5}.$$

Evaluating the product in parentheses,

$$\frac{12}{10} \times \frac{2}{3} = 5.$$

The fraction TWELVE-TENTHS solves the unknown-multiplier equation.

This last method, based on the multiplicative inverse, can be generalized: Let k, l, m and n be natural numbers, and consider the quotient

$$\frac{k}{m} \div \frac{l}{n},$$

whose value solves the unknown-multiplier equation,

$$\boxed{?} \times \frac{l}{n} = \frac{k}{m}.$$

Since

$$\frac{n}{\ell} \times \frac{\ell}{n} = 1,$$

it follows that

$$\frac{k}{m} \times \underbrace{\left(\frac{n}{l} \times \frac{l}{n}\right)}_{=1} = \frac{k}{m}.$$

Applying the associative property of multiplication on the left-half side,

$$\left(\frac{k}{m} \times \frac{n}{l}\right) \times \frac{l}{n} = \frac{k}{m}.$$

Evaluating the product in parentheses,

$$\frac{k \times n}{m \times l} \times \frac{l}{n} = \frac{k}{m}.$$

Comparing this equation with the unknown-multiplier equation, we finally obtain the infamous *invert-and-multiply* formula:

For every four natural numbers k, l, m and n

$$\frac{k}{m} \div \frac{l}{n} = \frac{k \times n}{m \times l}.$$

32.3 Misconceptions

32.3.1 Divided by two, divided by half

Consider the quotient,

$$5 \div \frac{1}{2}.$$

Its value, as we have just learned, is

$$5 \times \frac{2}{1} = 10.$$

More intuitively, this quotient represents the number of time that ONE-HALF of a whole is contained in 5 wholes; this number is indeed 10.

Yet, many children are likely to believe that this quotient equals TWO-AND-A-HALF. The reason behind this recurring error is that in spoken language, *divide by half* and *divide by two* are often used synonymously. One of the great challenge of mathematical instruction is to overcome misconceptions originating from interferences between mathematical terminology and colloquial language, as the latter is deeply rooted in our minds.

32.3.2 Is division a reducing operation?

On p. 252, we addressed the common misconception whereby multiplication is a magnifying operation. A similar misconception is common with respect to division: children (and adults) often perceive division as a reducing operation.

What is meant, when we say that division is reducing, is that the output of operators such as DIVIDE-BY FIVE is smaller than their input. However, just as an operator multiplying its input by a fixed number is not always magnifying, so an operator dividing its input by a fixed number is not always reducing. For example, the operator DIVIDE-BY ONE-HALF multiplies is input by 2, hence it is a magnifying operator.

The fact that not every division is reducing is counterintuitive. Children are often confused when evaluating expressions such as

$$5 \div \frac{1}{2} = 10,$$

discovering that the quotient, 10, is greater than the dividend, 5.

To conclude, the scaling properties of division are:

(a) Division by a number greater than 1 is reducing.
(b) Division by 1 does not change the input.
(c) Division by a number less than 1 is magnifying.

Mathematical problems

Problem 32.1. There are 40 snakes at the zoo, constituting TWO-THIRDS of all the reptiles. How many reptiles are there at the zoo?

(a) Write an appropriate unknown-factor equation.
(b) Write an appropriate arithmetic expression.
(c) Solve the problem using two different strategies.

Problem 32.2. Formulate four different word problems demonstrating in concrete terms the arithmetic expression

$$\frac{2}{3} \div \frac{4}{5}.$$

Problem 32.3. Which arithmetic expression corresponds to each of the following problems?

(a) A worker paves TWO-THIRDS of a sq-yard per hour. How many sq-yards will she pave in THREE-AND-A-HALF hours?
(b) A worker paves TWO-THIRDS of a sq-yard in THREE-AND-A-HALF hours. How many sq-yards does she pave per hour?
(c) A worker paves TWO-THIRDS of a sq-yard per hour. How long will it take her to pave THREE-AND-A-HALF sq-yards?
(d) A worker paves THREE-AND-A-HALF sq-yards per hour. How long will it take her to pave TWO-THIRDS of a sq-yard?

Problem 32.4. A snail crawled FIVE-AND-A-THIRD yards in THREE-QUARTERS of an hour. In this context, what does the following arithmetic expression mean,

$$5\frac{1}{3} \div \frac{3}{4}?$$

Similarly, what does the following arithmetic expression mean,

$$\frac{3}{4} \div 5\frac{1}{3}?$$

Hint: consider first the same question replacing the fractions by natural numbers.

Problem 32.5. Determine, without performing any calculation, which arithmetic expression is greater:

$$\frac{1}{2} \div \frac{1}{3} \quad \text{or} \quad \frac{2}{3} \div \frac{1}{4}?$$

Explain your answer.

Problem 32.6. Place the numbers 3, 4, 5 and 6 in the following template

$$\frac{\square}{\square} \div \frac{\square}{\square}$$

to obtain (a) a maximal quotient (b) a minimal quotient.

Problem 32.7. Fill-in the missing numbers on the following number line:

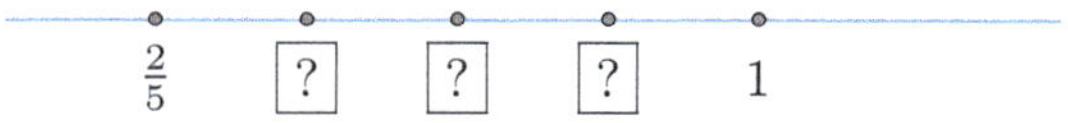

Pedagogical problems

Problem 32.8. Consider the following number sentence:

$$\frac{5}{8} \div \frac{2}{4} = \frac{5 \div 2}{8 \div 4} = \frac{2\frac{1}{2}}{2} = 1\frac{1}{4}.$$

Is it correct? How would you react to a child who uses this method for evaluating quotients of fractions?

Index

Common Core Index

www.ingramcontent.com/pod-product-compliance
Ingram Content Group UK Ltd.
Pitfield, Milton Keynes, MK11 3LW, UK
UKHW051423150926
8624IPUK00016B/576